An Anthology of
Military Quotations

By the same author

Internal Security Weapons and Equipment of the World
Brush Fire Wars: Campaigns of the British Army since 1945
The British Army in Northern Ireland
Northern Ireland Scrapbook
Weapons and Equipment of Counter Terrorism
The Art of Deception in Warfare
The Defence of the Nation

An Anthology of
Military Quotations

Edited by
MICHAEL DEWAR

ROBERT HALE · LONDON

ISBN 0 7090 3979 4

Robert Hale Limited
Clerkenwell House
Clerkenwell Green
London EC1R 0HT

Photoset in North Wales by
Derek Doyle & Associates, Mold, Clwyd.
Printed in Great Britain by
St Edmundsbury Press, Bury St Edmunds, Suffolk.
Bound by WBC Bookbinders Limited.

To Lavinia,
my long-suffering wife

Contents

The following was found in a pill box at Passchendaele in 1917 on its recapture by the Allies. The occupants had fought to the last man.

Special Orders to No. 1 Section

1. This position will be held and the section will remain here until relieved.

2. The enemy cannot be allowed to interfere with this programme.

3. If this section cannot remain here alive it will remain here dead but in any case it will remain here.

4. Should any man through shell-shock or such cause attempt to surrender he will stay here dead.

5. Should all guns be blown out the section will use Mills grenades and other novelties.

6. Finally the position as stated will be held.

Campbell Cpl

Acknowledgements

I should like to thank the staff of the Staff College Library at Camberley for their invaluable help in pointing me at the relevant books in my search for memorable quotations. Ian Drury of Aerospace Publishing was particularly generous in handing over to me his personal collection of quotations. But most important of all, my wife and two eldest sons, Alexander and James, managed to keep up a continuous bombardment of suitable quotations during the period I was compiling this volume. I am most grateful to them all.

I have made every effort to identify copyright holders and to obtain their permission, but would be glad to hear of any inadvertent errors or omissions.

Foreword

As an author I am constantly looking for good military quotations. As a schoolboy and at Sandhurst and at university I was always searching for apposite quotations with which to impress my tutors. I was still looking for a source of relevant military quotations at staff college ten years after leaving university. Max Hastings' *Oxford book of Military Anecdotes* is an excellent book but it fulfils a completely different purpose. It follows the chronological approach and quotes long passages from various books starting in biblical times and ending up in the present day. The choice of anecdotes is hugely entertaining but does not pretend to be comprehensive or indeed anything else other than haphazard. More important, the long anecdotes are not 'quotable quotes'.

My aim has been to produce a comprehensive compilation of famous quotations so as to provide a reference book for the military enthusiast, the student of military affairs and indeed anyone else who needs a source of apposite quotations for a military context. In many respects Max Hastings' volume and my own are complementary. Each certainly fulfils a completely separate purpose.

I agonized for a long time over the choice of the keywords in headings under which all the quotations could be classified. I could have included many more less obvious keywords and as many equally obscure quotations. I decided this would lessen the impact of the book, so I have tried, whilst remaining fairly comprehensive and catholic in my coverage, to include only those quotations which are evocative, hard hitting and memorable – in other words, 'quotable quotes'. It is also difficult to know precisely how to classify quotations: many fit easily under more than one keyword. In some instances quotations appear more than once if they are associated particularly strongly with both their author and another context. However, I have tried to keep this to a minimum in order to avoid unnecessary repetition.

Many readers will no doubt wonder why a particular quotation has not been included. The choice inevitably reflects my own interests, biases, prejudices and experience. Nevertheless, I hope this book will be a useful source of reference for military men and military enthusiasts for many years to come.

Michael Dewar

List of Keywords

Quotations

Action/Activity

Activity, activity, speed! (Activité, activité, vitesse!)
> Napoleon I, order to Massena before Eckmühl, 17 April 1809

A fundamental principle is never to remain completely passive, but to attack the enemy frontally and from the flanks, even while he is attacking us.
> Karl von Clausewitz, *Principles of War*, 1812

Activity in war is movement in a resistant medium. Just as a man immersed in water is unable to perform with ease and regularity the most natural and simplest movement, that of walking, so in war, with ordinary powers, one cannot keep even the line of mediocrity.
> Karl von Clausewitz, *On War*, 1832

It is even better to act quickly, and err than to hesitate until the time of action is past.
> Karl von Clausewitz, *On War*, 1832

We must make this campaign an exceedingly active one. Only thus can a weaker country cope with a stronger; it must make up in activity what it lacks in strength.
> Lieutenant-General Stonewall Jackson, letter, April 1863

Action is the governing rule of war.
> Ferdinand Foch, *Precepts*, 1919

A man who has to be convinced to act before he acts is not a
man of action ... You must act as you breathe.

Georges Clemençeau, 1841–1929, Army Staff College
papers

Adjutant

An adjutant is a wit ex officio.

Francis Grose, *Advice to the Officers of the British Army*,
1782

He was open-hearted, manly, friendly, an independent most
gallant and zealous officer, and much devoted to his own
corps. He neither cringed to, nor worshipped any man, but did
his duty manfully, and with impartiality – two qualities
inestimable in an adjutant.

Lieutenant Colonel J. Leech

Administration

Nothing shows a general's attention more than requiring a
number of returns, particularly such as it is difficult to make
with any degree of accuracy. Let your brigade-major, therefore,
make out a variety of forms, the more red lines the better: as to
the information they convey, that is immaterial; no one ever
reads them, the chief use of them being to keep the adjutants
and sergeants in employment.

Francis Grose, *Advice to the Officers of the British Army*,
1782

I have great difficulty in organising my commissariat for the
march and that department is very incompetent. The existence
of the army depends upon it and yet the people who manage it
are incapable of managing anything outside a country house.

The Duke of Wellington, Army Staff College papers

My Lord,

If I attempted to answer the mass of futile correspondence that surrounds me, I should be debarred from all serious business of campaigning. I must remind your Lordship – for the last time – that so long as I retain an independent position, I shall see that no officer under my Command is debarred, by attending to the futile drivelling of mere quill-driving in your Lordship's office, from attending to his first duty – which is, and always has been, so to train the private men under his command that they may, without question, beat any force opposed to them in the field.

> Attributed to the Duke of Wellington under date of 1810 but not to be found in any authentic source, this widely known quotation can in fact only be traced to a British Middle East Command training circular, c.1941, and is almost certainly spurious

It is very necessary to attend to detail, and to trace a biscuit from Lisbon into a man's mouth on the frontier, and to provide for its removal from place to place, by land and by water, or no military operations can be carried on.

> The Duke of Wellington, quoted in *Britain at Arms*, 1953

Napoleon directed Bourrienne to leave all his letters unopened for three weeks, and then observed with satisfaction how large a part of the correspondence had thus disposed of itself, and no longer required an answer.

> R.W. Emerson, *Representative Men*, 1850

> Stick close to your desks and never go to sea,
> And you all may be rulers of the Queen's Navee!
> W.S. Gilbert, HMS *Pinafore*, 1878

There has been a constant struggle on the part of the military element to keep the end – fighting, or readiness to fight – superior to mere administrative considerations ... The military man, having to do the fighting, considers that the chief necessity; the administrator equally naturally tends to think the smooth running of the machine the most admirable quality.

> Alfred Thayer Mahan, *Naval Administration and Warfare*, 1903

There is far too much paper in circulation in the Army, and no one can read even half of it intelligently.

Montgomery of Alamein, *Memoirs*, 1958

A man's ordinary day-to-day life must be well organised. Thus, hard conditions imposed on him in training to inculcate discipline do not rule out the desirability of good living quarters; and in the line a soldier's morale will be much improved if the administrative arrangements are good and if he is assured of proper conditions, with a reasonable amount of leisure and comfort when he leaves the front. But here a warning must be given. There is a danger today of 'welfare' being considered as an end in itself and not as a means to an end, one of the means of maintaining morale. Welfare by itself will not produce good morale because it is essentially soft; and it has already been stated that morale cannot be good unless it contains a quality of hardness. Hardness and privation are the school of the good soldier; idleness and luxury are his enemies.

Montgomery of Alamein, *Morale in Battle*, 1960s

In modern warfare no success is possible unless military units are adequately supplied with fuel, ammunition and food and their weapons and equipment are maintained. Modern battle is characterised by resolute and dynamic actions and by abrupt changes in the situation which call for a greater quantity of supplies than was the case during the Second World War. Hence the increasingly important role of logistic continuity aimed at supplying each soldier in good time with everything he needs for fulfilling his combat mission.

Col. Gen. Golushko, Chief of Logistic Staff, Soviet Armed Forces, 1984

Admiral

An admiral has to be put to death now and then to encourage the others [pour encourager les autres]

François-Marie Arouet Voltaire, *Candide*, 1759

Men go into the Navy ... thinking they will enjoy it. They do enjoy it for about a year, at least the stupid ones do, riding back and forth quite dully on ships. The bright ones find that they don't like it in half a year, but there's always the thought of that pension if only they stay in ... Gradually they become crazy. Crazier and crazier. Only the Navy has no way of distinguishing between the sane and the insane. Only about five per cent of the Royal Navy have the sea in their veins. They are the ones who become captains. Thereafter, they are segregated on their bridges. If they are not mad before this, they go mad then. And the maddest of these become admirals.

Attributed to George Bernard Shaw, 1856–1950

It is dangerous to meddle with admirals when they say they can't do things. They have always got the weather or fuel or something to argue about.

Sir Winston Churchill, to the Secretary of the US
Navy Frank Knox, December 1941

Advance

The duty of an advance guard does not consist in advancing or retiring, but in manoeuvring ... An advance guard should consist of picked troops, and the general officers, officers, and men should be selected for their respective capabilities and knowledge. An ill-trained unit is only an embarrassment to an advance guard.

Napoleon I, *Maxims of War*, 1831

When you move at night, without a light, in your own house, what do you do? Do you not (though it is a ground you know well) extend your arm in front so as to avoid knocking your head against the wall? The extended arm is nothing but an advance guard.

Ferdinand Foch, *Precepts*, 1919

The first duty of an advance guard is to advance.

Sir William Slim, *Unofficial History*, 1959

23

Agincourt

What's he that wishes so?
My cousin Westmoreland? No, my fair cousin:
If we are marked to die, we are now
To do our country loss: and if to live,
The fewer men, the greater share of honour.
God's will, I pray thee, wish not one man more.
By jove, I am not covetous for gold,
Nor care I who doth feed upon my cost:
It yearns me not if men my garments wear:
Such outward things dwell not in my desires.
But if it be a sin to covet honour,
I am the most offending soul alive.
No, faith, my coz, wish not a man from England:
God's peace, I would not lose so great an honour,
As one man more, methinks, would share from me,
For the best hope I have. O, do not wish one more:
Rather proclaim it, Westmoreland, through my host,
That he which hath no stomach to this fight,
Let him depart, his passport shall be made,
And crowns for convoy put into his purse:
We would not die in that man's company,
That fears his fellowship, to die with us.
This day is called the feast of Crispian:
He that outlives this day, and comes safe home,
Will stand a tip-toe when this day is named,
And rouse him at the name of Crispian.
He that outlives this day, and sees old age,
Will yearly on the vigil feast his neighbours,
And say, 'Tomorrow is Saint Crispian.'
Then will he strip his sleeve, and show his scars,
And say, 'These wounds I had on Crispin's day.'
Old men forget; yet all shall be forgot,
But he'll remember with advantages,
What feats he did that day. Then shall our names,
Familiar in his mouth as household words,
Harry the king, Bedford and Exeter,
Warwick and Talbot, Salisbury and Gloucester,
Be in their flowing cups freshly remembered.
This story shall the good man teach his son;
And Crispin Crispian shall ne'er go by,

From this day to the ending of the world,
But we in it shall be remembered;
We few, we happy few, we band of brothers,
For he today that sheds his blood with me
Shall be my brother: be he ne'er so vile,
This day shall gentle his condition.
And gentlemen in England, now a-bed,
Shall think themselves accursed they were not here:
And hold their manhoods cheap, whiles any speaks
That fought with us upon Saint Crispin's day.
>William Shakespeare, *King Henry V*, 1598

Air Force/Air Power

The nation that secures control of the air will ultimately control the world.
>Alexander Graham Bell, letter, 1909

That's good sport, but for the Army the plane is of no use.
>Ferdinand Foch, remark at the 1910 Circuit de l'Est, a plane race

The Independent Air Force should embody the greatest power compatible with the resources at our disposal; therefore no aerial resources should under any circumstances be diverted to secondary purposes, such as auxiliary aviation, local air defence, and anti-aircraft defence.
>Giulio Douhet, *The Command of the Air*, 1921

In order to assure an adequate national defence, it is necessary – and sufficient – to be in a position in case of war to conquer the command of the air.
>Giulio Douhet, *The Command of the Air*, 1921

I have mathematical certainty that the future will confirm my assertion that aerial warfare will be the most important element in future wars, and that in consequence not only will the importance of the independent Air Force rapidly increase, but the importance of the army and the navy will decrease in proportion.
>Giulio Douhet, *The Command of the Air*, 1921

Never in the field of human conflict was so much owed by so many to so few.

> Sir Winston Churchill, to the House of Commons, 20 August 1940 (of the RAF in the Battle of Britain)

The Navy can lose us the war, but only the Air Force can win it. Therefore, our supreme effort must be to gain overwhelming mastery in the air.

> Sir Winston Churchill, to the War Cabinet, 3 September 1940

The power of an air force is terrific when there is nothing to oppose it.

> Sir Winston Churchill, *The Gathering Storm*, 1948

Today air power is the dominant factor in war. It may not win a war by itself alone, but without it no major war can be won.

> Arthur Radford, speech, 1954

Modern air power has made the battlefield irrelevant.

> Sir John Slessor, *Strategy for the West*, 1954

Alliances/Allies

It is our true policy to steer clear of permanent alliances with any portion of the foreign world.

> George Washington, Farewell Address, 17 September 1796

It is a narrow policy to suppose that this country or that is to be marked out as the eternal ally or the perpetual enemy ... We have no eternal allies, and we have no eternal enemies. Our interests are eternal and perpetual, and those interests it is our duty to follow.

> Lord Palmerston, to the House of Commons, 1848

In war I would deal with the Devil and his grandmother.

> Joseph Stalin, 1879–1951, Army Staff College papers

Granting the same aggregate of force, it is never as great in two hands as in one, because it is not perfectly concentrated.

> Alfred Thayer Mahan, *Naval Strategy*, 1911

Any alliance whose purpose is not the intention to wage war is senseless and useless.

> Adolf Hitler, *Mein Kampf*, 1925

War without allies is bad enough – with allies it is hell!

> Sir John Slessor, *Strategy for the West*, 1954

Ambush

Those who wage war in mountains should never pass through defiles without first making themselves masters of the heights.

> Maurice de Saxe, *Reveries*, 1732

American Revolution

I rejoice that America has resisted. Three millions of people, so dead to all the feelings of liberty, as voluntarily to submit to be slaves, would have been fit instruments to make slaves of the rest.

> Lord Chatham (Pitt the Elder), to the House of Commons, 14 January 1766

Four or five frigates will do the business without any military force.

> Lord North, to the House of Commons, 1774

On this question of principle, while actual suffering was yet afar off [the American Colonies] raised their flag against a power, to which, for purposes of foreign conquest and subjugation, Rome, in the height of her glory, is not to be compared; a power which has dotted over the surface of the whole globe with her possessions and military posts, whose morning drum-beat, following the sun, and keeping company with the hours, circles the earth with one continuous and unbroken strain of the martial airs of England.

> Daniel Webster, to the Senate, 7 May 1834

The snow lies thick on Valley Forge,
The ice on Delaware,
But the poor dead soldiers of King George
They neither know nor care
>Rudyard Kipling, *The American Rebellion*, 1906

Ammunition

Put your trust in God, my boys, and keep your powder dry.
>Oliver Cromwell, to his troops at Marston Moor,
>2 July 1644

Amphibious Operations

When an advancing enemy crosses water do not meet him at
the water's edge. It is advantageous to allow half his force to
cross and then strike.
>Sun Tzu, 400–320 BC, *The Art of War*, chapter 9

The question of landing in face of an enemy is the most
complicated and difficult in war.
>Sir Ian Hamilton, *Gallipoli Diary*, 1920

Difficulties of landing on beaches are serious, even when the
invader has reached them; but difficulties of nourishing a
lodgement when exposed to heavy attack by land, air, and sea
are far greater.
>Sir Winston Churchill, note to the Chiefs of Staff
>Committee, 28 June 1940

A landing against organized and highly trained opposition is
probably the most difficult undertaking which military forces
are called upon to face.
>General George C. Marshall, during planning for the
>Sicilian landings, 1943

In landing operations, retreat is impossible. To surrender is as ignoble as it is foolish ... Above all else remember that we as the attackers have the initiative. We know exactly what we are going to do, while the enemy is ignorant of our intentions and can only parry our blows. We must retain this tremendous advantage by always attacking; rapidly, ruthlessly, viciously and without rest.

> Lieutenant-General George Patton, General Order to the Seventh Army before Sicily landings, 27 June 1943

We shall land at Inchon, and I shall crush them.

> Douglas MacArthur, to the Joint Chiefs of Staff, at Tokyo, 23 August 1950

The amphibious landing is the most powerful tool we have.

> Douglas MacArthur, planning conference for Inchon, Tokyo, 23 August 1950

Amphibious flexibility is the greatest strategic asset that a sea power possesses.

> B.H. Liddell Hart, *Deterrence or Defence*, 1960

Armed Forces

An armed, disciplined body is, in its essence, dangerous to liberty. Undisciplined, it is ruinous to society.

> Edmund Burke, speech on the Army Estimates, 1790

The services in war time are fit only for desperadoes, but in peace are fit only for fools.

> Benjamin Disraeli, *Vivian Grey*, 1827

... a class of men set apart from the general mass of the community, trained to particular uses, formed to peculiar notions, governed by peculiar laws, marked by peculiar distinctions.

> William Windham, 1750–1810, from speech to the House of Commons in 1807

What a society gets in its armed services is exactly what it asks for, no more and no less. What it asks for tends to be a reflection of what it is. When a country looks at its fighting forces it is looking in a mirror; the mirror is a true one and the face that it sees will be its own.

> General Sir John Hackett in *The Profession of Arms*, 1983

Armour

The Cavalry will never be scrapped to make room for the tanks; in the course of time Cavalry may be reduced as the supply of horses in this country diminishes. This depends greatly on the life of fox-hunting.

> *Journal of the Royal United Services Institute*, 1921

Where tanks are, is the front ... Wherever in future wars the battle is fought, tank troops will play the decisive role.

> Heinz Guderian, *Achtung! Panzer!* 1937

Armourer

The armourers, accomplishing the knights,
With busy hammers closing rivets up,
Give dreadful note of preparation.

> William Shakespeare, *King Henry V*, 1598

The first artificer of death; the shrewd,
Contriver who first sweated at the forge,
And forc'd the blunt and yet unbloodied steel
To a keen edge, and made it bright for war.

> William Cowper, *The Task*, 1758

Arms/Armament

Arms and the man I sing.
> Virgil, *Aeneid*, i, 19 BC

There cannot be good laws where there are not good arms.
> Niccolo Machiavelli, *The Prince*, 1513

When princes think more of luxury than of arms, they lose their state.
> Niccolo Machiavelli, *The Prince*, 1513

Perhaps more valid Armes,
Weapons more violent when next we meet,
May serve to better us, and worse our foes,
Or equal what between us made the odds,
In nature none ...
> John Milton, *Paradise Lost*, 1667

The subjects which are Protestants may have arms for their defence suitable to their conditions, and as allowed by law.
> The English Bill of Rights, December 1689

No glory is achieved except by arms. (Il n'y a pas de gloire achevée sans celle d'armes.)
> Marquis de Vauvenargues, *Works*, 1747

A well regulated Militia, being necessary to the security of a free State, the right of the people to keep and bear Arms, shall not be infringed.
> Constitution of the United States, Amendment II, 1791

The most solid moral qualities melt away under the effect of modern arms.
> Ferdinand Foch, *Precepts*, 1919

We can do without butter, but, despite all our love for peace, not without arms. One cannot shoot with butter but with guns.
> Paul Joseph Goebbels, speech in Berlin, 17 January 1936

It is war that shapes peace, and armament that shapes war.
> J.F.C. Fuller, *Armament and History*, 1945

31

It is customary in democratic countries to deplore expenditures on armaments as conflicting with the requirements of the social services. There is a tendency to forget that the most important social service that a government can do for its people is to keep them alive and free.

Sir John Slessor, *Strategy for the West*, 1954

Only when our arms are sufficient beyond doubt can we be certain that they will never be employed.

President John F. Kennedy, Inaugural Address, 20 January 1961

Army

To give a young gentleman right education.
The Army's the only good school in the nation.

Jonathan Swift, 1667–1745, *Hamilton Brown*

It is not big armies that win battles; it is the good ones.

Maurice de Saxe, *Reveries*, 1732

An army is composed for the most part of idle and inactive men, and unless the general has a constant eye upon them ... this artificial machine ... will very soon fall to pieces.

Frederick the Great, *Instructions to His Generals*, 1747

So sensible were the Romans of the imperfections of valour without skill and practice that, in their language, the name of an Army was borrowed from the word which signified exercise. ['Exercitus' = 'Army' in Latin.]

Edward Gibbon, *Decline and Fall of the Roman Empire*, 1776

The qualities which commonly make an army formidable are long habits of regularity, great exactness of discipline, and great confidence in the commander.

Samuel Johnson, 1709–84, Staff College papers

The country must have a large and efficient army, one capable of meeting the enemy abroad, or they must expect to meet him at home.

The Duke of Wellington, letter, 28 January 1811

The first measure for a country to adopt is to form an army.
> The Duke of Wellington, letter, 10 December 1811

I detest war. It spoils armies.
> Grand Duke Constantine of Russia, *c*.1820

The army is the people in uniform.
> Benjamin Constant, 1838–91

The army is a good book to open to study human life. There one learns to put his hand to everything, to the lowest and highest things. The most delicate and rich are forced to see poverty nearly everywhere, and to live with it, and to measure its morsel of bread and draught of water.
> Alfred de Bigny, *Military Service and Greatness*, 1835

The whole system of the army is something egregious and artificial. The civilian who lives out of it cannot understand it. It is not like other professions which require intelligence. A man one degree removed from idiocy, with brains sufficient to direct his power of mischief and endurance, may make a distinguished soldier.
> W.M. Thackeray (under pseudonym, 'Titmarsh'), 1811–63, in *Punch*

I have never seen so teachable and helpful a class as the Army generally.
> Florence Nightingale, letter to her sister, March 1856

The Army is the most outstanding institution in every country, for it alone makes possible the existence of all civic institutions.
> Helmuth von Moltke ('The Elder'), 1800–91, quoted in German Armed Forces Military History Institute papers, 1988

Governments needs armies to protect them against their enslaved and oppressed subjects.
> Leo Tolstoy, *The Kingdom of God is Within You*, 1893

It is my Royal and Imperial Command that you concentrate your energies, for the immediate present, upon one single purpose, and that is that you address all your skill and all the valour of my soldiers to exterminate first the treacherous English, and to walk over General French's contemptible little Army. [The origin of the nickname, 'Old Contemptibles' for the British Expeditionary Force in 1914]
> Kaiser Wilhelm II, General Order, Aix, 19 August 1914

The functions of an Army are: (1) to defeat the enemy's main force; (2) to seize upon his vitals.
> Sir Ian Hamilton, *The Soul and Body of an Army*, 1921

An army is a crowd – a homogeneous crowd, it is true, but retaining, despite its organization, some of the general characteristics of crowds: intense emotional suggestibility, obedience to leaders, etc. These factors must be handled by commanders.
> Gustave le Bon, *World in Revolt*, 1924

The Army should become a State within the State, but it should be merged in the State through service, in fact it should itself become the purest image of the State.
> General Hans von Seeckt, 1866–1936, *Thoughts of a Soldier*

History shows that there are no invincible armies.
> Joseph Stalin, address to the Russian people, 3 July 1941

An army is an institution not merely conservative but retrogressive by nature. It has such natural resistance to progress that it is always insured against the danger of being pushed ahead too fast. Far worse and more certain ... is the danger of it slipping backward. Like a man pushing a barrow uphill, if the soldier ceases to push, the military machine will run back and crush him.
> B.H. Liddell Hart, *Thoughts on War*, 1944

The Army, for all its good points, is a cramping place for a thinking man.
> B.H. Liddell Hart, *Thoughts on War*, 1944

The nature of armies is determined by the nature of the civilization in which they exist.

> B.H. Liddell Hart, 1895–1970, *The Ghost of Napoleon*

Army, British

The English never yield, and though driven back and thrown into confusion, they always return to the fight, thirsting for vengeance as long as they have a breath of life.

> Giovanni Mocenigo, to the Doge of Venice, 8 April 1588 (Mocenigo was Venetian Ambassador in Paris)

Of all the world's great heroes,
There's none that can compare
With a tow-row-row-row-row-row, for a British Grenadier!

> *The British Grenadiers*, attributed to Charles Dibdin and first performed 17 January 1780 in celebration of the attack and capture of Savannah (but a minority view dates words and music to the seventeenth century)

Ours [the British Army in the Peninsula] is composed of the scum of the earth – the mere scum of the earth. The British soldiers are fellows who have all enlisted for drink – that is the plain fact – they have all enlisted for drink.

> The Duke of Wellington, letter from Portugal, 1811

The British soldier can stand up to anything except the British War Office.

> George Bernard Shaw, *The Devil's Disciple*, 1897

The British Army should be a projectile to be fired by the British navy.

> Lord Grey, 1862–1933

War, therefore, will never cease, grievous though the thought may be. Yet, to descend again to lowly mundane things, its former outward manifestations seem likely to be transformed. It may well be that by new methods of scientific destruction the whole nature of armies may be changed. Infantry and Cavalry may vanish away and regiments and even armies in the old and honoured sense may cease to be. Then shall the British

Army likewise perish and its place shall know it no more. It matters not. Were the Army to be swept tomorrow into nothingness, it has already done enough to give it rank with the legions of ancient Rome. And it will be remembered best not for its surpassing valour and endurance, not for its countless deeds of daring and its invincible stubbornness in battle, but for its lenience in conquest and its gentleness in domination. Let Wellington's phrase be repeated once more: 'We are English and we pride ourselves on our deportment.'

Empires decay and fall and the British Empire cannot escape the common lot. Already the Dominions are virtually independent. They will forget, as the Americans have already forgotten, what they owe to the British soldier; but not the less will his work for them remain. In India the rule of the British will fade in due time into a legend of stolid white men, very terrible in fight, who swept the land from end to end, enforcing for a brief space strange maxims of equity and government. The age may be hereafter mournfully recalled by the Indian peasant as that wherein his forefathers reaped what they had sown under the protection of the British soldier. When the Empire shall have passed away it is the British soldier's figure that will loom out eminent above all, the calm upholder of the King's peace.

And the historian of the dim future, summing up the whole story, may conclude it in such words as these. 'The builders of this Empire despised and derided the stone which became the headstone of this corner. They were not worthy of such an Army. Two centuries of persecution could not wear out its patience; two centuries of thankless toil could not abate its ardour; two centuries of conquest could not awake it to insolence. Dutiful to its masters, merciful to its enemies, it clung steadfastly to its old simple ideals – obedience, service, sacrifice.'

Sir John Fortescue, 1859–1933, *A History of The British Army*

No incident is more familiar in our military history than the stubborn resistance of the British line at Waterloo. Through the long hours of the midsummer day, silent and immovable the squares and squadrons stood in the trampled corn, harassed by an almost incessant fire of cannon and of musketry, to which they were forbidden to make reply. Not a moment but heard some cry of agony; not a moment but some comrade fell headlong in the furrows. Yet as the bullets of the skirmishers hailed around them, and the great round shot tore through the tight-packed ranks, the word was passed quietly. 'Close in on the centre, men'; and as the sun neared his setting, the regiments, still shoulder to shoulder, stood fast upon the ground they had held at noon. The spectacle is characteristic. In good fortune and in ill it is rare indeed that a British regiment does not hold together; and this indestructible cohesion, best of all the qualities that an armed body can possess, is based not merely on hereditary resolution, but on mutual confidence and mutual respect. The man in the ranks has implicit faith in his officer, the officer an almost unbounded belief in the valour and discipline of his men;

> Colonel Henderson, *The Science of War*, 1906

We had this ... If ever an army fought in a just cause we did. We coveted no man's country; we wished to impose no form of government on any nation. We fought for the clean, the decent, the free things of life, for the right to live our lives in our own way, as others could live theirs, to worship God in what faith we chose, to be free in body and mind, and for our children to be free. We fought only because the powers of evil had attacked these things ...

> Sir William Slim, *Defeat into Victory*, 1956

Arnhem (17 September 1944)

'Not in vain' may be the pride of those who survived and the epitaph of those who fell.

> Sir Winston Churchill, to the House of Commons, 28 September 1944

Arsenal

We must be the great arsenal of Democracy.

> President Franklin D. Roosevelt, address to the American People, 29 December 1940

Artillery

And if the Turks by means of their artillery gained the victory over the Persians and the Egyptians, it resulted from no other merit than the unusual noise, which frightened the cavalry ... artillery is useful to an army when the soldiers are animated by the same valour as that of the ancient Romans, but without that it is perfectly inefficient, especially against courageous troops.
> Niccolo Machiavelli, *Discourses*, 1531

It is with artillery that war is made.
> Napoleon I, after Löbau, May 1809

The best generals are those who have served in the artillery.
> Napoleon I, to General Gaspard Gourgaud, St Helena, 1815

Leave the artillerymen alone. They are an obstinate lot.
> Napoleon I, 1769–1821, British Army Artillery papers

Artillery, like the other arms, must be collected in mass if one wishes to attain a decisive result.
> Napoleon I, 1769–1821

The better the infantry, the more it should be economized and supported by good batteries. Good infantry is without doubt the sinews of an army; but if it has to fight a long time against very superior artillery, it will become demoralized and will be destroyed.
> Napoleon I, *Maxims of War*, 1831

I have seen war, and faced modern artillery, and I know what an outrage it is against simple men.
> T.M. Kettle, *The Ways of War*, 1915

Our thousand-stringed artillery began to play its battle-tune.
> Field Marshal Paul Von Hindenburg, *Out of My Life*, 1920

It is of great value to an army, whether in defence or offence, to have at its disposal a mass of heavy batteries.
> Sir Winston Churchill, *The Gathering Storm*, 1948

Once, from the safety of a well-dug command post, I looked down on a battery of artillery in action in the African bush. It was firing at five rounds per gun per minute and, idly I timed the nearest gun. In that area the enemy, unfortunately, had complete local air supremacy, and guns, unless engaged in some vital task, were ordered to remain silent, whenever hostile aircraft appeared. Gradually, dominating all other sound, came the dull drone of bombers, flying low; but the guns went on firing, five rounds per gun per minute, for they were supporting an infantry attack. The first stick of bombs fell around the gun I was watching. Some of its crew were hit. The dry bush roared into flame, which spread instantly to the camouflage nets over the gun. It vanished from my sight in smoke and flame. Yet from the very midst of that inferno, at the exact intervals, came the flash and thud of the gun firing. Five rounds per gun per minute. Never a falter, never a second out. No weak link there; discipline held.

Sir William Slim, *Courage and Other Broadcasts*, 1951

Assault

Once more into the breach, dear friends, once more;
Or close the wall up with our English dead!
William Shakespeare, *King Henry V*, 1598

On no account should we overlook the moral effect of a rapid, running assault. It hardens the advancing soldier against danger, while the stationary soldier loses his presence of mind.
Karl von Clausewitz, *Principles of War*, 1812

Atlantic, Battle of (1941–5)

The battle of the Atlantic was the dominating factor all through the war. Never for one moment could we forget that everything happening elsewhere ... depended ultimately on its outcome.
Sir Winston Churchill,: *Closing the Ring*, 1951

Attack

Decline the attack altogether unless you can make it with advantage.

> Maurice de Saxe, *Reveries*, 1732

I approve of all methods of attacking provided they are directed at the point where the enemy's army is weakest and where the terrain favours them the least.

> Frederick the Great, *Instructions for His Generals*, 1747

Gentlemen, the enemy stands behind his entrenchments, armed to the teeth. We must attack him and win, or else perish. Nobody must think of getting through any other way. If you don't like this, you may resign and go home.

> Frederick the Great, to his officers, before the battle of Leuthen, 5 December 1757

But, in case Signals can neither be seen or perfectly understood, no Captain can do very wrong if he places his Ship alongside that of an Enemy.

> Horatio Nelson, plan of attack before Trafalgar, 9 October 1805

Up Guards, and at them!

> Attributed to the Duke of Wellington, as his command for counterattack by the Brigade of Guards at Waterloo, 18 June 1815

I was too weak to defend, so I attacked.

> Robert E. Lee (attributed), 1807–70, Confederate General, American Civil War

For what is more thrilling than the sudden and swift development of an attack at dawn?

> Sir Winston Churchill, *The River War*, 1899

Attack, whatever happens! ... Victory will come to the side that outlasts the other.

> Ferdinand Foch, Order during the Battle of the Marne, 7 September 1914

Hard pressed on my right. My centre is yielding. Impossible to manoeuvre. Situation excellent. I am attacking.

> Ferdinand Foch, message to Marshal Joffre, battle of the Marne, 8 September 1914

A well conducted battle is a decisive attack successfully carried out.

> Ferdinand Foch, *Precepts*, 1919

Strength lies not in defence but in attack.

> Adolf Hitler, *Mein Kampf*, 1925

We are so outnumbered there's only one thing to do. We must attack.

> Sir Andrew Browne Cunningham, before attacking the Italian fleet at Taranto, 11 November 1940

The Victor will be the one who finds within himself the resolution to attack; the side with only defence is inevitably doomed to defeat.

> M. Frunze, Soviet military theorist, twentieth century

Audacity

Impetuosity and audacity often achieve what ordinary means fail to achieve.

> Niccolo Machiavelli, *Discourses*, 1531

Arm me, audacity, from head to foot.

> William Shakespeare, *Cymbeline*, 1609

Audacity, audacity again, and audacity always! (De l'audace, encore de l'audace, et toujours de l'audace!)

> Georges Danton, to the French Legislative Assembly, 2 September 1792

In audacity and obstinacy will be found safety.

> Napoleon I, *Maxims of War*, 1831

If the theory of war does advise anything, it is the nature of war to advise the most decisive, that is the most audacious.

> Karl von Clausewitz, *Principles of War*, 1812

Never forget that no military leader has ever become great without audacity.

Karl von Clausewitz, *Principles of War*, 1812

My critics ... want war too methodical, too measured; I would make it brisk, bold, impetuous, perhaps sometimes even audacious.

Antoine Henri Jomini, *Summary of the Art of War*, 1838

Aviation, Army

That's good sport, but for the Army the aeroplane is of no use.

Ferdinand Foch, remark at the 1910 'Circuit of the East' Air Race

The surprise ambush is a particular danger to forward or outflanking Soviet detachments, operating in isolation with open flanks, or to a battalion which has experienced artillery fire and will therefore be closed down and looking to its front. It will not see a flank attack by helicopters.

Christopher Donnelly, lectures, 1985

Balaklava (25 October 1854)

It is magnificent, but it is not war.
> Pierre Bosquet, on observing the charge of the Light
> Brigade, 25 October 1854

Half a league, half a league,
Half a league onward,
All in the valley of death
Rode the six hundred.
> Alfred Lord Tennyson, *The Charge of the Light Brigade*,
> 1854

Bands

I am delighted at the action you have taken about bands, but when are we going to hear them playing about the streets? Even quite small parade marches are highly beneficial ... In fact, wherever there are troops and leisure for it there should be an attempt at military display.
> Sir Winston Churchill, note to Secretary of State for
> War, 12 July 1940

Bases

Ships ... must have secure ports to which to return, and must be followed by the protection of their country throughout the voyage.
> Alfred Thayer Mahan, *The Influence of Sea Power upon
> History*, 1890

Important naval stations should be secured against attack by land as well as by sea.

Alfred Thayer Mahan, *Naval Strategy*, 1911

Battalion

A battalion is made up of individuals, the least important of whom may chance to delay things or somehow make them go wrong.

Karl von Clausewitz, *On War*

Battle, Battlefield

The race is not to the swift, nor the battle to the strong.

Ecclesiastes 9

So ends the bloody business of the day.

Homer, *c*.1000 BC, *Odyssey*

I do not favour battles, particularly at the beginning of a war. I am sure a good general can make war all his life and not be compelled to fight one.

Maurice de Saxe, *Reveries*, 1732

When we enter the lists of battle, we quit the sure domain of truth and leave the decision to the caprice of chance.

William Godwin, *An Enquiry Concerning Political Justice*, 1793

The business of an English Commander-in-Chief being first to bring an Enemy's Fleet to Battle on the most advantageous terms to himself (I mean that of laying his Ships close on board the Enemy, as expeditiously as possible); and secondly to continue them there until the Business is decided ...

Horatio Nelson, excerpt from Order to the Fleet, 1805

... my precise object ... a close and decisive Battle.

Horatio Nelson, excerpt from Order to the Fleet, 1805

Battles decide everything.
> Karl von Clausewitz, *Principles of War*, 1812

Between a battle lost and a battle won, the distance is immense and there stand empires.
> Napoleon I, on the eve of the Battle of Leipzig, 15 October 1813

A battle sometimes decides everything; and sometimes the most trifling thing decides the fate of a battle.
> Napoleon I, letter to Barry E. O'Meara, St Helena, 9 November 1816

I hope to God I have fought my last battle. It is a bad thing to be always fighting. While in the thick of it I am too much occupied to feel anything; but it is wretched just after. It is quite impossible to think of glory.
> The Duke of Wellington, to Lady Frances Shelley after Waterloo, Brussels, 19 June 1815

I always say that, next to a battle lost, the greatest misery is a battle gained.
> Attributed to the Duke of Wellington by Frances, Lady Shelley, *Diary*

The battle may therefore be regarded as War concentrated, as the centre of effort of the whole war or campaign. As the sun's rays unite in the focus of a concave mirror in a perfect image, and in the fulness of their heat; so the forces and circumstances of war unite in a focus in the great battle for one concentrated utmost effort.
> Karl von Clausewitz, *On War*, 1832

So all day long the noise of battle roll'd
Among the mountains by the winter sea,
Until King Arthur's table, man by man,
Had fallen in Lyonnesse about their lord.
> Alfred Lord Tennyson, *The Passing of Arthur*, 1842

Where a battle has been fought, you will find nothing but the bones of men and beasts: where a battle is being fought, there are hearts beating.
> Henry David Thoreau, 1817–62

I have seen battles, too –
Have waded foremost in their bloody waves,
And heard their hollow roar of dying men.
> Matthew Arnold, *Sohrab and Rustum*, 1853

Battle is the ultimate to which the whole life's labour of an officer should be directed. He may live to the age of retirement without seeing a battle; still, he must always be getting ready for it as if he knew the hour and the day it is to break upon him. And then, whether it come late or early, he must be willing to fight – he must fight.
> Brigadier General C.F. Smith, US Army, to Colonel Lew Wallace, September 1861

In a larger sense we cannot dedicate, we cannot hallow this ground. The brave men, living and dead, who struggled here, have consecrated it far above our poor power to add or detract.
> President Abraham Lincoln, Gettysburg Address, 19 November 1863

Read here the moral roundly writ
For him who into battle goes –
Each soul that, hitting hard or hit,
Endureth gross or ghostly foes,
Prince, blown by many overthrows,
Half blind with shame, half choked with dirt,
Man cannot tell, but Allah knows
How much the other side was hurt!
> Rudyard Kipling, *Verses on Games*, 1898

Modern battle ... is a struggle between nations, fighting for their existence, for independence, or for some less noble interest; fighting, anyhow, with all their resources and all their passions. These masses of men and of passions have to be shaken and overthrown.
> Ferdinand Foch, *Precepts*, 1919

Battles are won by slaughter and manoeuvre. The greater the general, the more he contributes in manoeuvre, the less he demands in slaughter.
> Sir Winston Churchill, *The World Crisis*, vol.II, 1923

We love battle. If battle should at length die out of the world, then all joy would die out of life.

> Busso Loewe, *Creed of the German Pagan Movement*,
> 1936

France has lost a battle. But France has not lost the war.

> President Charles de Gaulle, broadcast to the French
> People, 18 June 1940

Battle is the most magnificent competition in which a human being can indulge. It brings out all that is best; it removes all that is base.

> Lieutenant-General George Patton, to officers, 45th
> Division, before the Sicily landings, 27 June 1943

Battle should no longer resemble a bludgeon fight, but should be a test of skill, a manoeuvre combat, in which is fulfilled the great principle of surprise by striking 'from an unexpected direction against an unguarded spot'.

> B.H. Liddell Hart, *Thoughts on War*, 1944

The late M. Venizelos observed that in all her wars England – he should have said Britain, of course – always wins one battle – the last.

> Sir Winston Churchill, speech at the Lord Mayor's
> Luncheon, London, 10 November 1942

While the battles the British fight may differ in the widest possible ways, they have invariably two common characteristics – they are always fought uphill and always at the junction of two or more map sheets.

> Sir William Slim, *Unofficial History*, 1959

Battle Experience

War is a singular art. I assure you that I have fought sixty battles, and I learned nothing but what I knew when I fought the first one.

> Napoleon I, to General Gaspard Gourgaud,
> St Helena, 1815

I love a brave soldier who has undergone the baptism of fire, whatever nation he may belong to.
> Napoleon I, to Barry O'Meara, St Helena, 1816

Use makes a better soldier than the most urgent considerations of duty – familiarity with danger enabling him to estimate the danger. He sees how much is the risk, and is not afflicted with imagination; knows practically Marshal Saxe's rule, that every soldier killed costs the enemy his weight in lead.
> R.W. Emerson, 1803–82, *Leavenworth Papers*

Nothing is more exhilarating than to be shot at without result.
> Sir Winston Churchill, *The Malakand Field Force*, 1898

Bayonet

Rangers of Connaught! It is not my intention to expend any powder this evening. We'll do this business with the cold iron!
> Sir Thomas Picton, to the 88th Foot before the assault on Badajoz, 6 April 1812

Blitzkrieg

We had seen a perfect specimen of the modern Blitzkrieg; the close interaction on the battlefield of army and air force; the violent bombardment of all communicatons and of any town that seemed an attractive target; the arming of an active Fifth Column; the free use of spies and parachutists; and above all, the irresistible forward thrust of great masses of armour.
> Sir Winston Churchill, *The Gathering Storm*, 1948

Blockade

Methought the hindering of their trade the best provocation to make the enemy's fleet come out.
> Lord Sandwich, after outbreak of the Second Dutch War, 1655

A battle is really nothing to the fatigue and anxiety of such life as we lead. It is now thirteen months since I let go an anchor, and, from what I see, it may be as much longer.

> Cuthbert Collingwood, letter while on blockade
> station off Cadiz, 1806

Blockade in order to be binding must be effective.

> Declaration of Paris, Art. 4, 1856

... that most hopeless form of hostilities, an inadequate commercial blockade and a war on seaborne trade.

> Sir Julian Corbett, *The Successors of Drake*, 1900

It [blockade] is a belligerent measure which touches every member of the hostile community, and, by thus distributing the evils of war, as insurance distributes the burden of other losses, it brings them home to every man.

> Alfred Thayer Mahan, *Some Neglected Aspects of War*,
> 1907

Blood

What coast knows not our blood? (Quae caret ora cruore nostro?)

> Horace, *Odes*, ii, 1, 23 BC

Blood is the god of war's rich livery.

> Christopher Marlowe, *Tamburlaine the Great*, 1587

The purple testament of bleeding war.

> William Shakespeare, *King Richard II*, 1596

Blood is the price of victory.

> Karl von Clausewitz, *On War*, 1832

Not by speechifying and counting majorities are the great questions of the time to be solved – that was the error of 1848 and 1849 – but by iron and blood.

> Otto von Bismarck, to the Prussian Diet, 30
> September 1862

If blood be the price of admiralty, Lord God, we ha' paid in full!

> Rudyard Kipling, *The Song of the English*, 1893

Historical experience is written in blood and iron.

> Mao Tse-tung, *On Guerrilla Warfare*, 1937

I would say to the House, as I said to those who have joined this Government, 'I have nothing to offer but blood, toil, tears and sweat'.

> Sir Winston Churchill to the House of Commons, 13 May 1940

Blunder

Some one had blundered.

> Alfred Lord Tennyson, *The Charge of the Light Brigade*, 1854

Boldness

The gods favour the bold.

> Ovid, *Metamorphoses*, x, *c.* AD 5

Great empires are not maintained by timidity.

> Tacitus, *Histories*, *c.* AD 115.

Be bold, be bold, and everywhere be bold.

> Edmund Spenser, *The Faerie Queene*, 1609

Boldness be my friend!
Arm me, audacity, from head to foot!

> William Shakespeare, *Cymbeline*, 1609

The measure may be thought bold, but I am of opinion the boldest are the safest.

> Horatio Nelson, to Sir Hyde Parker, urging immediate, vigorous action against the Danes and Russians, 24 March 1801

Desperate affairs, require desperate remedies.

> Horatio Nelson, 1758–1805

Perhaps I should not insist on this bold manoeuvre, but it is my style, my way of doing things.
> Napoleon I, letter to Prince Eugene, 1813

Bold decisions give the best promise of success,
> Field Marshal Erwin Rommel, 1891–1944, *Rules of Desert Warfare*

'Safety first' is the road to ruin in war.
> Sir Winston Churchill, telegram to Anthony Eden, 3 November 1940

Brave/Bravery

Fortune favours the brave. (Fortes fortuna adiuvat.)
> Terence, *Phormio*, c.160 BC

Few men are born brave; many become so through training and force of discipline.
> Vegetius, *The Military Institutions of the Romans*, iii, AD 378

Few men are brave by nature, but good order and experience make many so. Good order and discipline in any army are more to be depended upon than courage alone.
> Niccolo Machiavelli, *Art of War*, 1520

What's brave, what's noble,
Let's do it after the high Roman fashion,
And make death proud to take us.
> William Shakespeare, *Antony and Cleopatra*, 1606

Brave men are brave from the first blow.
> Pierre Corneille, *The Cid*, 1636

A brave man never dies.
> Owen Feltham, d. 1688, *Resolves ('Of Fame')*

None but the brave deserves the fair.
> John Dryden, *Alexander's Feast*, 1697

Women are partial to the brave, and they think every man handsome who is going to the camp or the gallows.

John Gay, *The Beggar's Opera*, 1728

He [Chevalier Folard] supposes all men to be brave at all times and does not realize that the courage of the troops must be reborn daily, that nothing is so variable, and that the true skill of the general consists in knowing how to guarantee it.

Maurice de Saxe, *Reveries*, 1732

The brave man is not he who feels no fear,
For that were stupid and irrational;
But he, whose noble soul its fear subdues,
And bravely shares the danger nature shrinks from.

Joanna Baillie, 1762–1851, *Basil*

That man is not truly brave who is afraid either to seem or to be, when it suits him, a coward.

Edgar Allan Poe, *Marginalia*, 1844–9

As to the way in which some of our Ensigns and Lieutenants braved danger – the boys just come out of school – it exceeds all belief. They ran as at cricket!

The Duke of Wellington in Samuel Rogers, *Recollections*, 1859

Bravery never goes out of fashion.

William Makepeace Thackeray, *The Four Georges*, 1860

At the bottom of a good deal of the bravery that appears in the world there lurks a miserable cowardice. Men will face powder and steel because they cannot face public opinion.

George Chapin, US commentator, 1826–80

The bravest are the tenderest,
The loving are the daring.

Bayard Taylor, *The Song of the Camp*, 1864

Britain, Battle of (1940)

Let us therefore brace ourselves to our duties, and so bear ourselves that, if the British Empire and its Commonwealth last for a thousand years, men will say, 'This was their finest hour.'

Sir Winston Churchill, to the House of Commons,
18 June 1940

Far out on the grey waters of the North Sea and the Channel coursed and patrolled the faithful, eager flotillas peering through the night. High in the air soared the fighter pilots, or waited serene at a moment's notice around their excellent machines. This was a time when it was equally good to live or die.

Sir Winston Churchill, *Their Finest Hour*, 1949

The Bombs have shattered my churches,
have torn my streets apart,
But they have not bent my spirit
and they shall not break my heart.
For people's faith and courage
are lights of London town
Which still would shine in legends though
my last broad bridge were down.

Greta Briggs, *London Under Bombardment*, 1941

Bugle

One blast upon his bugle horn
Were worth a thousand men.

Walter Scott, *Lady of the Lake*, 1810

Blow, bugle, blow; set the wild echoes flying
Blow, bugle, blow; answer, echoes, dying, dying, dying.

Alfred Lord Tennyson, *The Princess*, 1850

What are the bugles blowin' for said
    Files-on-Parade.
'To turn you out, to turn you out,' the
    Colour-Sergeant said.

Rudyard Kipling, *Danny Deever*, 1890

53

Blow out, you bugles, over the rich dead ...
Rupert Brooke, *The Dead*, 1914

Built-up Areas, Fighting in

What is the position about London? I have a very clear view that we should fight every inch of it, and that it would devour quite a large invading army.
Sir Winston Churchill, memorandum to General Ismay, 2 July 1940

Urban warfare is regarded as an exception, an occasional and unhappy accident, far away from the main stream. War, when properly conducted, according to human superstition, belongs in civilisationless open countryside.
General S.L.A. Marshall, *Notes for Urban Warfare*

Bullets

O you leaden messengers,
That ride upon the violent speed of fire,
Fly with false aim.
William Shakespeare, *All's Well That Ends Well*, 1602

That shall be my music in the future!
Charles XII of Sweden, on first hearing the whistle of bullets in battle, Copenhagen, August 1700

I heard the bullets whistle; and believe me, there is something charming in the sound.
George Washington, letter to his mother after the battle of Great Meadows, 3 May 1754

Every bullet hath its billet.
John Wesley, *Journal*, 6 June 1765

The flying bullet down the Pass,
That whistles clear, 'All flesh is grass.'
Rudyard Kipling, *Arithmetic on the Frontier*, 1886

There is nothing more democratic than a bullet or a splinter of steel.

> Wendell Willkie, *An American Programme*, 1944

Bureaucracy

The British bureaucrat has managed to transform inertia from a negative into a positive force. Bureaucracy is one huge 'sit tight' club.

> Sir Ian Hamilton, *The Soul and Body of an Army*, 1921

Burial

With all respect and rites of burial.
Within my tent his bones tonight shall lie,
Most like a soldier, order'd honourably.
> William Shakespeare, *Julius Caesar, 1599*

I'll hide my master from the flies, as deep
As these poor pick-axes can dig.
> William Shakespeare, *Cymbeline*, 1609

… her own clay shall cover, heaped and pent,
Rider and horse – friend, foe – in one red burial blent!
> George, Lord Byron, *Childe Harold's Pilgrimage* (The Eve of Waterloo), 1816

Not a drum was heard, not a funeral note,
As his corse to the rampart we hurried;
Not a soldier discharged his funeral shot
O'er the grave where our hero we buried.
> Charles Wolfe: *The Burial of Sir John Moore at Corunna*, 1817

In Flanders fields the poppies blow
Between the crosses, row on row.
> John McRae, *In Flanders Fields*, 1917

Caesar, Caius Julius (102–44 BC)

I came, I saw, I conquered. (Veni, vidi, vici.)
>Despatch to the Roman Senate after the battle of Zela, 47 BC

When Caesar says 'Do this', it is perform'd.
>William Shakespeare, *Julius Caesar*, 1599

Camouflage

Let every soldier hew him down a bough and bear't before
    him: thereby shall be shadow
The numbers of our host, and make discovery
Err in report of us.
>William Shakespeare, *Macbeth*, 1605

Camp, Encampment

The hum of either army stilly sounds,
That the fixed sentinels almost receive
The secret whispers of each other's watch;
Fire answers fire, and through their paly flames
Each battle sees the other's umbered face;
Steed threatens steed, in high and boastful neighs
Piercing the night's dull ear; and from the tents
The armourers, accomplishing the knights,
With busy hammers closing rivets up,
Give dreadful note of preparation.
>William Shakespeare, *King Henry V*, 1598

The first rule is always to occupy the heights; the second, that if
you have a river or a stream in front of the camp, not to move
more than half a musket-shot's distance from it.
>   Frederick the Great, *Instructions for His Generals*, 1747

The art of encamping in position is the same as taking up the
line in order of battle in this position. To this end, the artillery
should be advantageously placed, ground be selected which is
not commanded or liable to be turned, and, as far as possible,
the guns should command and cover the surrounding country.
>   Napoleon I, *Maxims of War*, 1831

Encampments of the same army should always be formed so as
to protect each other.
>   Napoleon I, *Maxims of War*, 1831

Cannon

Cannon and fire-arms are cruel and damnable machines; I
believe them to have been the direct suggestion of the Devil.
>   Martin Luther, *Table Talk*, 1569

By East and West let France and England mount.
Their battering cannon charged to the mouths.
>   William Shakespeare, *King John*, 1596

The cannons have their bowels full of wrath.
And ready mounted are they to spit forth
Their iron indignation.
>   William Shakespeare, *King John*, 1596

And nearer, clearer, deadlier than before!
Arm! Arm! it is – it is – the cannon's opening roar!
>   George, Lord Byron, *Childe Harold's Pilgrimage* (The
>   Eve of Waterloo), 1816

The cannon's breath
Wings far the hissing globe of death.
>   George, Lord Byron, 1788–1824, *The Siege of Corinth*

Cannon to the right of them,
Cannon to the left of them,
Cannon in front of them
Volley'd and thunder'd
> Alfred Lord Tennyson, *The Charge of the Light Brigade*,
> 1854

I heard the hoarse-voiced cannon roar,
The red-mouthed orators of war.
> Joaquin Miller, 1841–1913, US author

Cannon his name,
Cannon his voice, he came.
> George Meredith, *Napoleon*, 1891

Capitulation

All generals, officers, and soldiers who capitulate in battle to save their own lives should be decimated. He who gives the order and those who obey are alike traitors and deserve capital punishment.
> Napoleon I, *Maxims of War*, 1831

Captain

A brave captain is a root, out of which, as branches, the courage of his soldiers doth spring.
> Sir Philip Sidney, 1554–86

Who can look for modestie and sobrietie in the souldiers, where the captaine is given to wine, or women, and spendeth his time in riot and excesse?
> Mathew Sutcliffe, *The Practice, Proceedings, and Lawes of Armes*, 1593

That in the captain's but a choleric word
Which in the soldier is flat blasphemy.
> William Shakespeare, *Measure for Measure*, 1604

I take a bold Step, a rakish Toss, a smart Cock, and an impudent Air to be the principal ingredients in the Composition of a Captain.

> George Farquhar, *The Recruiting Officer*, 1706

The union of wise theory with great character will constitute the great captain.

> Antoine Henri Jomini, *Summary of the Art of War*, 1838

The captain, in the first place, is lord paramount. He stands no watch, comes and goes when he pleases, and is accountable to no one, and must be obeyed in everything.

> Richard Dana, *Two Years Before the Mast*, 1840

Casualties

How are the mighty fallen in the midst of battle.

> II Samuel 1

There are few that die well in a battle.

> William Shakespeare, *King Henry V*, 1598

Every individual nobly did his duty; and it is observed that our dead … were lying, as they fought, in ranks, and every wound was in the front.

> Sir William Beresford, report after Albuera, 16 May 1811

You could not be successful in such an action without a large loss. We must make up our minds to affairs of this kind sometimes, or give up the game.

> The Duke of Wellington, letter to Sir William Beresford, 19 May 1811, after Albuera

I had never heard of a battle in which everybody was killed; but this seemed likely to be an exception, as all were going by turns.

> Captain John Kincaid, fl. 1815, *Adventures with the Rifle Brigade* (reminiscence of Waterloo)

Let us not hear of generals who conquer without bloodshed. If bloody slaughter is a horrible sight, then that is a ground for paying more respect to war.

Karl von Clausewitz, *On War*, 1832

'You know,' he [the White Knight] said gravely, 'it's one of the most serious things that can possibly happen to one in a battle – to get one's head cut off!'

C.L. Dodgson (Lewis Carroll), *Through the Looking Glass*, 1872

Casualties? What do I care about casualties?

Attributed to Major General A.G. Hunter-Weston, Gallipoli, 1915

They claim that to fire human grapeshot at the enemy without preparation, gives us moral ascendancy. But the thousands of dead Frenchmen, lying in front of the German trenches, are instead those who are giving moral ascendancy to the enemy.

Abel Ferry, memorandum to the Cabinet of French Premier Viviani, 1916

The most fatal heresy in war, and, with us, the most rank, is the heresy, that battles can be won without heavy loss.

Sir Ian Hamilton, *Gallipoli Diary*, 1920

Pile the bodies high at Austerlitz and Waterloo,
Shovel them under and let me work –
I am the grass; I cover all.

Carl Sandburg, 1878–1964, *Grass*

A big butcher's bill is not necessarily evidence of good tactics.

Sir Archibald Wavell, reply to Churchill, August 1940, when reproached for having evacuated British Somaliland with only 260 casualties

Causes of War

Wars spring from unseen and generally insignificant causes, the first outbreak being often but an explosion of anger.

Thucydides, *History of the Peloponnesian War*, ii, *c.*404 BC

In war, important events result from trivial causes.
Julius Caesar, *The Gallic Wars*, 51 BC

The same reasons that make us quarrel with a neighbour cause war between two princes.
Michel de Montaigne, *Essays*, 1580

In the nature of man we find three principal causes of quarrel. First, competition; secondly, diffidence; thirdly, glory. The first maketh men invade for gain; the second, for safety; the third, for reputation.
Thomas Hobbes, *Leviathan*, 1651

The human heart is the starting-point in all matters pertaining to war.
Maurice de Saxe, *Reveries*, 1732

Vice foments war; virtue fights. Were there no virtue, we should have peace forever.
Marquis de Vauvenargues, *Works*, 1747

Each government accuses the other of perfidy, intrigue and ambition, as a means of heating the imagination of their respective nations, and incensing them to hostilities. Man is not the enemy of man, but through the medium of a false system of government.
Thomas Paine, *The Rights of Man*, 1791

The well-spring of war is in the human heart.
Stephen Luce, 1827–1917

States which are weak from a military standpoint, and which are surrounded by stronger neighbours invite war, and if they neglect their military organizations from false motives, they court this danger by their own supineness.
Colmar von der Goltz, *Nation in Arms*, 1883

War is the usual condition of Europe. A thirty years' supply of causes of war is always on hand.
Prince Kropotkin, *Speeches from a Revolution*, 1884

When a war is waged by two opposing groups of robbers for the sake of deciding who shall have a freer hand to oppress more people, then the question of the origin of the war is of no real economic or political significance.
>
> V.I. Lenin, article in *Pravda*, 26 April 1917

Is there any man here or any woman – let me say, is there any child – who does not know that the seed of war in the modern world is industrial and commercial rivalry?
>
> Woodrow Wilson, speech in St Louis, 5 September 1919

It takes at least two to make a peace, but one can make a war.
>
> Neville Chamberlain, speech in Birmingham, 28 January 1939

The urge to gain release from tension by action is a precipitating cause of war.
>
> B.H. Liddell Hart, *Thoughts on War*, 1944

Cavalry

The horse is the strength of the army. The horse is a moving bulwark.
>
> *The Hitopadesa*, III, *c.* AD 500

Ten thousand cavalry only amount to ten thousand men. No one has ever died in battle through being bitten or kicked by a horse. It is men who do whatever gets done in battle.
>
> Xenophon, speech to the Greek army after the defeat of Cyrus at Cunaxa, 401 BC

Altogether, cavalry operations are exceedingly difficult, knowledge of the country is absolutely necessary, and ability to comprehend the situation at a glance, and an audacious spirit, are everything.
>
> Maurice de Saxe, *Reveries*, 1732

Whereever gentlemen can hunt, there can cavalry act.
> Sir Banastre Tarleton, to the House of Commons,
> 1810

And there was mounting in hot haste, the steed,
The mustering squadron, and the clattering car,
Went pouring forward with impetuous speed,
And swiftly forming in the ranks of war.
> George, Lord Byron, *Childe Harold's Pilgrimage* (Eve of
> Waterloo), 1816

Some enthusiasts today talk about the probability of the horse becoming extinct and prophesy that the aeroplane, the tank and the motor-car will supersede the horse in future wars ... I am sure that as time goes on you will find just as much use for the horse – the well bred horse – as you have done in the past.
> Sir Douglas Haig, interview, 1925

Ceremonial

Bad luck to this marching,
Pipe-claying and starching;
How neat one must be to be killed by the French!
> Soldiers' song, Peninsular War, 1811

What makes a regiment of soldiers a more noble object of view than the same mass of mob? Their arms, their dress, their banners, and their art and artificial symmetry of their position and movements.
> George, Lord Byron, letter to John Murray,
> 7 February 1821

Wherever there are troops and leisure for it, there should be an attempt at military display.
> Sir Winston Churchill, note to Secretary of State for
> War, 12 July 1940

No one who has participated in it or seen it well done should doubt the inspiration of ceremonial drill.
> Field Marshal Earl Wavell, *Soldiers and Soldiering*, 1953

Chance

Now good or bad, 'tis but the chance of war.
> William Shakespeare, *Troilus and Cressida*, 1601

We should make war without leaving anything to chance, and in this especially consists the talent of a general.
> Maurice de Saxe, *Reveries*, 1732

Something must be left to chance; nothing is sure in a Sea Fight beyond all others.
> Horatio Nelson, Plan of Attack, before Trafalgar, 9 October 1805

There is no human affair which stands so constantly and so generally in close connection with chance as War.
> Karl von Clausewitz, *On War*, 1832

The affairs of war, like the destiny of battles, as well as of empires, hang upon a spider's thread.
> Napoleon I, *Political Aphorisms*, 1848

Countless and inestimable are the chances of war.
> Sir Winston Churchill, *The River War*, 1899

Change of Command

To remove a General in the midst of a campaign – that is the mortal stroke.
> Duke of Marlborough, 1650–1722

When promoted to the command of a regiment from some other corps, show them that they were all in the dark before, and, overturning their whole routine of discipline, introduce another as different as possible ... If you can only contrive to vamp up some old exploded system, it will have all the appearance of novelty to those, who have never practised it before: the few who have, will give you credit for having seen a great deal of service.
> Francis Grose, *Advice to the Officers of the British Army*, 1782

Charge

Charge, and give no foot of ground.
> William Shakespeare, *III King Henry VI*, 1590

I see you stand like greyhounds in the slips.
Straining upon the start. The game's afoot:
Follow your spirit; and, upon this charge
Cry 'God for Harry! England and Saint George!'
> William Shakespeare, *King Henry V*, 1598

Chariot

For thou shalt drive out the Canaanites, though they have iron
chariots, and though they be strong.
> Joshua 17

Thy walls shall shake at the noise of the horsemen, and of the
wheels, and of the chariots.
> Ezekiel 26

Bring me my bow of burning gold!
Bring me my arrows of desire!
Bring me my spear! O clouds unfold!
Bring me my chariot of fire!
> William Blake, 1757–1827, *Jerusalem*

Chivalry

I shall maintain and defend the honest adoes and quarrels of all
ladies of honour, widows, orphans, and maids of good fame.
> The Oath of Knighthood (written out by Sir William
> Drummond of Hawthornden, 1619)

None but the brave deserves the fair.
> John Dryden, *Alexander's Feast*, 1697

The age of chivalry is gone; and that of sophisters, economists, and calculators, has succeeded.

> Edmund Burke, *Reflections on the Revolution in France*, 1790

So faithful in love, and so dauntless in war,
There never was knight like the young Lochinvar.

> Walter Scott, *Marmion* (Lochinvar), 1808

As a soldier, preferring loyal and chivalrous warfare to organized assassination if it be necessary to make a choice, I acknowledge that my prejudices are in favour of the good old times when the French and English Guards courteously invited each other to fire first – as at Fontenoy – preferring them to the frightful epoch when priests, women and children throughout Spain plotted the murder of isolated soldiers.

> Antoine Henri Jomini, *Summary of the Art of War*, 1838

Generals think war should be waged like the tourneys of the Middle Ages. I have no use for knights. I need revolutionaries.

> Adolf Hitler, to Hermann Rauschning, 1940

Christian Soldier

A good Christian will never make a bad soldier.

> Gustavus Adolphus, after the landing at Usedom, 24 June 1630

If you choose godly, honest men to be captains of Horse, honest men will follow them.

> Oliver Cromwell, letter to Sir William Springer, September 1643

Onward, Christian soldiers
Marching as to war,
With the cross of Jesus
Going on before!

> Sabine Baring-Gould, hymn, *Onward Christian Soldiers*, 1864

We are on our last legs, owing to the delay in the expedition. However, God rules all, and as He will rule to His glory and our welfare, His will be done ... I am quite happy, thank God, and like Lawrence, I have 'tried to do my duty'.

> Charles George Gordon, last letter from Khartoum, 14 December 1884

If the historian of the future should deem my service worthy of some slight reference, it would be my hope that he mention me not as a commander engaged in campaigns and battles, even though victorious to American arms, but rather as one whose sacred duty it became, once the guns were silenced, to carry to the land of our vanquished foe the solace and hope and faith of Christian morals.

> General Douglas MacArthur, 1880–1964

Churchill, Sir Winston (1874–1965)

Winston is back.

> General message from the Admiralty to all ships and shore stations, 3 September 1939, announcing Churchill's reappointment as First Lord of the Admiralty, the post he had held in World War I

Civil War

There is nothing unhappier than a civil war, for the conquered are destroyed by, and the conquerors destroy, their friends.

> Dionysius of Halicarnassus, *Antiquities of Rome*, c. 20 BC

Civil dissension is a viperous worm.

> William Shakespeare, *I King Henry VI*, 1591

The war of all against all.

> Thomas Hobbes, *The Citizen*, 1642

In such battles [civil wars] those who win are established as loyalists, the vanquished as traitors.
Pedro Calderon, 1600–81

Civil wars strike deepest of all into the manners of the people. They vitiate their politics; they corrupt their morals; they pervert even the natural taste and relish of equity and justice. By teaching us to consider our fellow-citizens in a hostile light, the whole body of our nation becomes gradually less dear to us.
Edmund Burke, letter to the Sheriffs of Bristol, 3 April 1777

I am one of those who have probably passed a longer period of my life engaged in war than most men, and principally in civil war; and I must say this, that if I could avoid, by any sacrifice whatever, even one month of civil war in the country to which I was attached, I would sacrifice my life in order to do it.
Duke of Wellington, to the House of Lords, March 1829

A foreign war is a scratch on the arm; a civil war is an ulcer which devours the vitals of a nation.
Victor Hugo, *Ninety-Three*, 1879

Civilians

In all cases the military should be under strict subordination to and governed by the civil power.
Virginia Declaration of Rights, 12 June 1776

Though soldiers are the true supports,
The natural allies of courts,
Woe to the Monarch who depends
Too much on his red-coated friends;
For even soldiers sometimes think –
Nay, Colonels have been known to reason –
And reasoners, whether clad in pink,
Or Red, or blue, are on the brink
(Nine cases out of ten) of treason.
Thomas Moore, 1779–1852

We shall wait in vain for the awakening in our country of that public spirit which the English and the French and other peoples possess, if we do not imitate them in setting for our military leaders certain bounds and limitations which they must not disregard.

Carl Zuckmayer, *Germany, A Final Prologue*, 1896

An army cannot be directed in war nor commanded in peace under the immediate authority of a civilian. There must be a military commander, the obedient servant of the Government supported by the Government in the exercise of his powers ... and sheltered by the Government against all such criticism as would weaken his authority.

Spencer Wilkinson, *The Brain of an Army*, preface to 1913 edition

What soldier relishes the sight of a civilian flourishig a sword?

Philip Guedalla, *Wellington*, 1931

Civilians have had no instruction, training or experience in the principles of war, and to that extent are complete amateurs in the methods of waging war. It is idle however, to pretend that intelligent men whose minds are concentrated for years on one task learn nothing about it by daily contact with its difficulties and the way to overcome them.

Lloyd George, *War Memoirs*, 1937

Our arms must be subject to ultimate civilian control and command at all times, in war as well as peace. The basic decisions on our participation in any conflict and our response to any threat – including all decisions relating to the use of nuclear weapons, or the escalation of a small war into a large one – will be made by the regularly constituted civilian authorites.

President John F. Kennedy, 1917–63, to the House of Representatives.

Colonels

There are no bad regiments; there are only bad colonels.

Napoleon I, 1769–1821

For soldiers sometimes think –
Nay, Colonels have been known to reason ...
Thomas Moore, 1779–1852

Colours

The soldiers should make it an article of faith never to abandon
their standard. It should be sacred to them; it should be
respected; and every type of ceremony should be used to make
it respected and precious.
Maurice de Saxe, *Reveries*, 1732

Stood for his country's glory fast,
And nail'd her colours to the mast!
Sir Walter Scott, *Marmion, 1808*

Every means should be taken to attach the soldier to his
colours.
Napoleon I, *Maxims of War*, 1831

Command

By command I mean the general's qualities of wisdom,
sincerity, humanity, courage, and strictness.
Sun Tzu, 400–320 BC, *The Art of War*, chapter 1

When on active service the commander must prove himself
conspicuously careful in the matter of forage, quarters,
water-supply, outposts, and all other requisites; forecasting the
future and keeping ever a wakeful eye in the interest of those
under him; and in case of any advantage won, the truest gain
which the head of affairs can reap is to share with his men the
profits of success.

Indeed, to put the matter in a nutshell, there is small risk a
general will be regarded with contempt by those he leads, if,
whatever he may have to preach, he shows himself best able to
perform.

If, further, the men shall see in their commander, one who, with the knowledge how to act, has force of will and cunning to make them get the better of the enemy; and, if, further, they have the notion well into their heads that this same leader may be trusted not to lead them recklessly against the foe, without the help of Heaven, or despite the auspices – I say, you have a list of virtues which make those under his command the more obedient to their ruler.

> Xenophon, c.400 BC quoted in Staff College papers

I am a man under authority, having soldiers under me: and I say to this man, Go, and he goeth; and to another, Come, and he cometh.

> Matthew 8

A prince should therefore have no other aim or thought, nor take up any other thing for his study, but war and its organization and discipline, for that is the only art that is necessary to one who commands.

> Niccolo Machiavelli, *The Prince*, 1513

He who wishes to be obeyed must know how to command.

> Niccolo Machiavelli, *Discourses*, 1531

The success of my whole project is founded on the firmness of the conduct of the officer who will command it.

> Frederick the Great, *Instructions for His Generals*, 1747

If you are deficient in knowledge of your duty, the word of command given in a boatswain's tone of voice, with a tolerable assurance ... will carry you through till you get a smattering of your business.

> Francis Grose, *Advice to the Officers of the British Army*, 1782

I can no longer obey; I have tasted command, and I cannot give it up.

> Napoleon I, in conversation with Miot de Melito, 1798

I command – or I hold my tongue.

> Napoleon I, *Political Aphorisms*, 1848

Nothing is so important in war as an undivided command.

> Napoleon I, *Maxims of War*, 1831

There are Field Marshals who would not have shone at the head of a cavalry regiment, and vice versa.
Karl von Clausewitz, *On War*, 1832

The best means of organizing the command of an army ... is to: (1) Give the command to a man of tried bravery, bold in the fight and of unshaken firmness in danger. (2) Assign as his chief of staff a man of high ability, of open and faithful character, between whom and the commander there may be perfect harmony.
Antoine Henri Jomini, *Summary of the Art of War*, 1838

The power to command has never meant the power to remain mysterious.
Ferdinand Foch, *Precepts*, 1919

In action it is better to order than to ask.
Sir Ian Hamilton, *Gallipoli Diary*, 1920

There is required for the composition of a great commander not only massive common sense and reasoning power, not only imagination, but also an element of legerdemain, an original and sinister touch, which leaves the enemy puzzled as well as beaten.
Sir Winston Churchill, *The World Crisis*, vol.II, 1923

The high commands of the Army are not a club. It is my duty and that of His Majesty's Government to make sure that exceptionally able men, even though not popular with their military contemporaries, should not be prevented from giving their services to the Crown.
Sir Winston Churchill, note for the Secretary of State for War, 4 September 1942

A commander must accustom his staff to a high tempo from the outset, and continually keep them up to it. If he once allows himself to be satisfied with norms, or anything less than an all-out effort, he gives up the race from the starting post, and will sooner or later be taught a bitter lesson.
Field Marshal Erwin Rommel, 1891–1944

I do not play at war. I shall not allow myself to be ordered about by commanders-in-chief. I shall make war. I shall determine the correct moment for attack. I shall shrink from nothing.
Adolf Hitler, *Regarding the Russian Campaign*, 1943

Command doth make actors of us all.
> John Masters, *The Road Past Mandalay*, 1961

Commander

The commander of an army is one in whom civil and martial acumen are combined. To unite resolution with resilience is the business of war.
> Wu Ch'i, 430–381 BC, *Art of War*

The troops are scattered, and the commanders very poor
  rogues.
> William Shakespeare, *All's Well That Ends Well*, 1602

The Commander of an Army neither requires to be a learned explorer of history nor a publicist, but he must be well versed in the higher affairs of state; he must know and be able to judge correctly of traditional tendencies, interests at stake, the immediate questions at issue, and the characters of leading persons; he need not be a close observer of men, a sharp dissector of human character, but he must know the character, the feelings, the habits, the peculiar faults and inclinations, of those whom he is to command. He need not understand anything about ... the harness of a battery horse, but he must know how to calculate exactly the march of a column ... These are matters only to be gained by the exercise of an accurate judgement in the observation of things and men.
> Karl von Clausewitz, *On War*, 1832

Our army would be invincible if it could be properly organized and officered. There were never such men in an army before. They will go anywhere and do anything if properly led. But there is the difficulty – proper commanders.
> R.E. Lee, to Stonewall Jackson, 1862

The military commander is the fate of the nation.
> Helmuth von Moltke ('Moltke the Elder'), 1800–91

Great results in war are due to the commander. History is

therefore right in making generals responsible for victories – in which case they are glorified; and for defeats – in which case they are disgraced.

Ferdinand Foch, *Precepts*, 1919

A commander should have a profound understanding of human nature, the knack of smoothing out troubles, the power of winning affection while communicating energy, and the capacity for ruthless determination where required by circumstances. He needs to generate an electrifying current, and to keep a cool head in applying it.

B.H. Liddell Hart, *Thoughts on War*, 1944

The commander must be the prime mover of the battle and the troops must always have to reckon with his appearance in personal control.

Erwin Rommel, *The Rommel Papers*, 1953

There is one quality above all which seems to me essential for a good commander, the ability to express himself clearly, confidently, and concisely, in speech and on paper ... It is a rare quality amongst Army Officers, to which not nearly enough attention is paid in their education. It is one which can be acquired, but seldom is, because it is seldom taught.

Field Marshal Earl Wavell, *Soldiers and Soldiering*, 1953

One of the most valuable qualities of a commander is a flair for putting himself in the right place at the vital time.

Sir William Slim, *Unofficial History*, 1959

The commander's will must rest on iron faith: faith in God, in his cause, or in himself.

Correlli Barnett, *The Swordbearers*, 1963

Communications

The Book of Military Administration says: 'As the voice cannot be heard in battle, drums and bells are used. As troops cannot see each other clearly in battle, flags and banners are used.' Now gongs and drums, banners and flags are used to focus the attention of the troops. When the troops can be thus united, the

brave shall not advance alone, nor shall the cowardly withdraw. This is the art of employing a host.
Sun Tzu 400–320 BC

An army must have but one line of operations. This must be maintained with care and abandoned only for major reasons.
Napoleon I, *Maxims of War*, 1831

Communications dominate war; broadly considered, they are the most important single element in strategy, political or military.
Alfred Thayer Mahan, *The Problem of Asia*, 1900

Comradeship

All military organizations, land or sea, are ultimately dependent upon open communications with the basis of national power.
Alfred Thayer Mahan, *Naval Strategy*, 1911

It is not for myself, or on my own account chiefly, that I feel the sting of disappointment. No! It is for my brave officers, for my noble-minded friends and comrades, such a gallant set of fellows! Such a band of brothers! My heart swells at the thought of them!
Horatio Nelson, 1758–1805

Concentration

It is better to be on hand with ten men than to be absent with ten thousand.
Tamerlane, 1336–1405, Staff College papers.

If you meet two enemies, do not each attack one. Combine both on one of the enemy; you will make sure of that one, and you may also get the other afterwards; but, whether the other escape or not, your country will have won a victory, and gained a ship.
Horatio Nelson, 1758–1805, Instructions to two frigate captains sent off on detached service

When you have resolved to fight a battle, collect your whole force. Dispense with nothing. A single battalion sometimes decides the day.

Napoleon I, *Maxims of War*, 1831

In any military scheme that comes before you, let your first question to yourself be, Is this consistent with the requirement of concentration?

Alfred Thayer Mahan, *Naval Strategy*, 1911

The principles of war could, for brevity, be condensed into a single word – 'Concentration'.

B.H. Liddell Hart, *Thoughts on War*, 1944

An army should always be so distributed that its parts can aid each other and combine to produce the maximum possible concentration of force at one place, while the minimum of force necessary is used elsewhere to prepare the success for the concentration.

B.H. Liddell Hart in *Strategy: the Indirect Approach*, 1929

To concentrate all is an unrealizable ideal, and dangerous even as a hyperbole. Moreover, in practice the 'minimum necessary' may form a far larger proportion of the total than the 'maximum possible'.

B.H. Liddell Hart in *Strategy: the Indirect Approach*, 1929

Conflict

Never in the field of human conflict was so much owed by so many to so few.

Sir Winston Churchill, *Hansard*, 20 Aug 1940

Conquer, Conquest

In the practical art of war, the best thing of all is to take the enemy's country whole and intact; to shatter and destroy it is not so good ... Hence, to fight and conquer in all your battles is not supreme excellence; supreme excellence consists in breaking the enemy's resistance without fighting.

> Sun Tzu, 100–320 BC, *The Art of War*, Chapter 3

I came, I saw, I conquered. (Veni, vidi, vici.)

> Julius Caesar, despatch to the Roman Senate after the battle of Zela, 47 BC

Whoever conquers a free town and does not demolish it commits a great error and may expect to be ruined himself.

> Niccolo Machiavelli, *The Prince*, 1513

The English conquered us, but they are far from being our equals.

> Napoleon I, letter to General Gaspard Gourgaud, St Helena, 1815

Conquest is the most debilitating fever of a nation and the rudest of glories ... The sword does not plough deep.

> Francis Lieber, 1800–72

Conscription

Conscription is the vitality of a nation, the purification of its morality, and the real foundation of all its habits.

> Napoleon 1, *Political Aphorisms*, 1848

The system of conscription has always tended to foster quantity at the expense of quality.

> B.H. Liddell Hart, *The Untimeliness of a Conscript Army*, 1950

All the countries which collapsed under the shock of the German blitzkrieg in 1939, 1940, and 1941 relied on long-established conscript armies for their defence.

> B.H. Liddell Hart, *Defence of the West*, 1950

Copenhagen, Battle of (2 April 1801)

It is warm work; and this day may be the last to any of us at a moment. But mark you! I would not be elsewhere for thousands.

> Horatio Nelson, at the height of action against the
> Danish fleet at Copenhagen, 2 April 1801

Of Nelson and the North
Sing the glorious day's renown.

> Thomas Campbell, 1777–1844, *Battle of the Baltic*

Counter Attack

We have retired far enough for today; you know I always sleep upon the field of battle!

> Napoleon I, to retreating French troops before the
> counter stroke at Marengo, 14 June 1800

When you are occupying a position which the enemy threatens to surround, collect all your force immediately, and menace him with an offensive movement.

> Napoleon I, *Maxims of War*, 1831

Counterattack is the soul of defence. Defence is in a passive attitude, for that is the negation of war. Rightly conceived it is attitude of alert expectation. We wait for the moment when the enemy shall expose himself to a counterstroke, the success of which will so far cripple him as to render us relatively strong enough to pass to the offensive ourselves.

> Julian Corbett, *Some Principles of Maritime Strategy*,
> 1911

To every blow struck in war there is a counter.

> Sir Winston Churchill, memorandum for the War
> Cabinet, 16 December 1939

Courage

The strongest, most generous, and proudest of all virtues is true courage.
> Michel de Montaigne, *Essays*, 1580

'Tis true, that we are in great danger;
The greater therefore should our courage be.
> William Shakespeare, *King Henry V*, 1598

Courage is a quality so necessary for maintaining virtue that it is always expected, even when it is associated with vice.
> Samuel Johnson, 1709–84

Intelligence alone is not courage. We often see the most intelligent people are irresolute.
> Karl von Clausewitz, *On War*, 1832

If courage is the first characteristic of the soldier, perseverance is the second.
> Henri Plon, *Correspondence of Napoleon I*, 1863

There are only two classes who, as categories, show courage in war – the front-line soldier and the conscientious objector.
> B.H. Liddell Hart, *Thoughts on War*, 1944

Courage is rightly esteemed the first of human qualities ... because it is the quality which guarantees all others.
> Sir Winston Churchill, 1874–1965

There is nothing like seeing the other fellow run to bring back your courage.
> Sir William Slim, *Unofficial History*, 1959

Coward

Cowards die many times before their deaths;
The valiant never taste of death but once.
> William Shakespeare, *Julius Caesar*, 1599

It is vain for the coward to fly; death follows close behind; it is by defying it that the brave escape.

François-Marie Arouet Voltaire, 1694–1778

It is mutual cowardice that keeps us in peace. Were one-half of mankind brave, and one-half cowards, the brave would be always beating the cowards. Were all brave, they would lead a very uneasy life; all would be continually fighting; but being all cowards, we go on very well.

Samuel Johnson, in Boswell's *Life of Johnson*, 28 April 1778

I could not look on Death, which being known.
Men led me to him, blindfold and alone.

Rudyard Kipling, *Epitaphs of the War (The Coward)*, 1919

Danger

Out of this nettle, danger, we pluck this flower, safety.
William Shakespeare, *I King Henry IV*, 1597

To conquer without danger is to triumph without glory. (A vaincre sans peril, on triomphe sans glorie.)
Pierre Corneille, *The Cid*, 1637

Dangers, by being despised, grow great; so they do by absurd provision against them.
Edmund Burke, speech on the Petition of the Unitarians, 11 May 1792

If I had been censured everytime I have run my ship, or fleets under my command, into great danger, I should long ago have been out of the Service, and never in the House of Peers.
Horatio Nelson, letter to the Admiralty, March 1805

Death, Dead

It is sweet and fitting to die for one's country. (Dulce et decorum est pro patria mori).
Horace, 65–8 BC, *Odes*, iii, 2

We who are about to die, salute you. (Morituri te salutamus.)
Traditional greeting of the gladiators to Roman Emperors

Those that leave their valiant bones in France,
Dying like men, though buried in your dunghills,
They shall be famed.
William Shakespeare, *King Henry V*, 1598

Nothing in his life
Became him like the leaving of it; he died
As one that studied in his death
To throw away the dearest thing he ow'd
As 'twere a careless trifle.

William Shakespeare, Macbeth, 1605

O eloquent, just and mightie DEATH! whom none could advise, thou has perswaded; what none hath dared, thou hast done; and whom all the world hath flattered, thou only hast cast out of the world and despised; thou has drawn together all the farre stretched greatnesse, all the pride, cruelltie, and ambition of men, and covered it all with these two narrow words, HIC JACET!

Sir Walter Raleigh: Historie of the World, 1615

Once more unto the breach, dear friends,
   Once more;
Or close the wall up with our English dead!
In peace there's nothing so becomes a man
As modest stillness and humility:
But when the blast of war blows in our ears,
Then imitate the action of the tiger;
Stiffen the sinews, summon up the blood,
Disguise fair nature with hard-favour'd rage.

William Shakespeare, King Henry V, 1598

So he passed over, and all the trumpets sounded for him on the other side.

John Bunyan, Pilgrim's Progress (Death of Valiant for Truth), 1678

Home they brought her warrior dead.
She nor swoon'd, nor utter'd cry:
All her maidens, watching said,
'She must weep or she will die.'

Alfred Lord Tennyson, The Princess, 1847

Dead on the field of honour. (Mort au champs d'honneur)

Theophile de la Tour d'Auvergne, whom Napoleon called 'First Grenadier of France,' was killed at, Oberhausen 27 June 1800. Thereafter, at reveille his name was answered with the above words

Better like Hector in the field to die,
Than like a perfumed Paris turn and fly.
>   Henry Wadsworth Longfellow: *Morituri Salutamus*,
>   1875

We have fed our sea for a thousand years
And she calls us, still unfed,
Though there's never a wave of all her waves
But marks our English dead.
>   Rudyard Kipling, *A Song of the English*, 1893

Happy are those who have died in great battles,
Lying on the ground before the face of God.
>   Charles Peguy, *c*.1910

They shall not grow old, as we that are left grow old:
Age shall not weary them nor the years condemn.
>   Laurence Binyon, *For the Fallen*, 21 September 1914

Blow out, you bugles, over the rich dead!
There's none of these so lonely and poor of old,
But, dying, has made us rarer gifts than gold.
>   Rupert Brooke, *The Dead*, 1914

If I should die, think only this of me:
That there's some corner of a foreign field
That is forever England. There should be
In that rich earth a richer dust concealed,
A dust whom England bore, shaped, made aware,
gave, once, her flowers to love, her ways to roam,
A body of England's, breathing English air,
washed by the rivers, blest by suns of home.
And think, this heart, all evil shed away,
A pulse in the eternal wind, no less
Gives somewhere back the thoughts by England given;
Her sights and sounds; dreams happy as her day;
And laughter, learnt of friends; and gentleness
In hearts at peace, under an English heaven
>   Rupert Brooke, 1887–1915, *The Soldier*

When Spring comes back with rustling shade
And apple blossoms fill the air

I have a rendezvous with Death
When Spring brings back blue days and fair.

> Alan Seeger, *I Have a Rendezvous with Death*, 1916

Red lips are not so red.
As the stained stones kissed by the English dead.

> Wilfred Owen, 1893–1918, *Greater Love*

Here dead lie we because we did not choose
To live and shame the land from which we sprung.
Life, to be sure, is nothing much to lose;
But young men think it is, and we were young.

> A.E. Housman, *Here Dead Lie We*, 1922

It is better to die on your feet than to live on your knees.

> La Pasionaria (Dolores Ibarruri), speech in Paris, 1936

Death stands at attention, obedient, expectant, ready to serve, ready to shear away the peoples en masse; ready, if called on, to pulverize, without hope of repair, what is left of civilisation. He awaits only the word of command. He awaits it from a frail, bewildered being, long his victim, now – for one occasion only – his Master.

> Sir Winston Churchill, *The Gathering Storm*, 1948

Deception

All warfare is based on deception. Therefore, when capable, feign incapacity; when active, inactivity. When near, make it appear that you are far away; when far away that you are near. Offer the enemy a bait to lure him; feign disorder and strike him.

> Sun Tzu, 400–320 BC, *The Art of War*, i

Though fraud in other activities be detestable, in the management of war it is laudable and glorious, and he who

overcomes an enemy by fraud is as much to be praised as he
who does so by force.

> Niccolo Machiavelli, *Discourses*, 1531

Force and fraud, are in war the two cardinal virtues.

> Thomas Hobbes, *Leviathan*, 1651

To achieve victory we must as far as possible make the enemy
blind and deaf by sealing his eyes and ears, and drive his
commanders to distraction by creating confusion in their
minds.

> Mao Tse-tung, 1893–1976, *On Protracted War*

Ruses, such as dummy positions, should be employed to
misdirect the enemy, waste his ammunition, and attract him
into funnels in which he can be enfiladed.

> B.H. Liddell Hart, *Thoughts on War*, 1920

In wartime truth is so precious that she should always be
attended by a bodyguard of lies.

> Attributed to Sir Winston Churchill

Decision

The wine is poured and we must drink it.

> Marshal Ney, order to advance, Jena, 14 October 1906

You will usually find that the enemy has three courses open to
him, and of these he will adopt the fourth.

> Helmuth von Moltke ('The Elder'), 1800–91

In all operations a moment arrives when brave decisions have
to be made if an enterprise is to be carried through.

> Sir Roger Keyes, *Letter from the Dardanelles*, 1915

When all is said and done the greatest quality required in a
commander is 'decision'; he must be able to issue clear orders
and have the drive to get things done. Indecision and
hesitation are fatal in any officer; in a C-in-C they are criminal.

> Montgomery of Alamein, *Memoirs*, 1958

Decorations

Show me a republic, ancient or modern, in which there have been no decorations. Some people call them baubles. Well, it is by such baubles that one leads men.

> Napoleon I, remark on introducing the Legion of Honour, 19 May 1802

A soldier will fight long and hard for a bit of coloured ribbon.

> Napoleon I, to the Captain, *HMS Bellerophon*, 15 July 1815

Orders and decorations are necessary in order to dazzle the people.

> Napoleon I, *Political Aphorisms*, 1848

The General got 'is decorations thick
  (The men that backed 'is lies could not complain),
The Staff 'ad DSO's till we was sick,
An' the soldier 'ad to do the work again!

> Rudyard Kipling, *Stellenbosch*, 1903

Defeat

Even the final decision of a war is not to be regarded as absolute. The conquered nation often sees it as only a passing evil, to be repaired in after times by political combinations.

> Karl von Clausewitz, *On War*, 1832

We are not interested in the possibilities of defeat. They do not exist.

> Queen Victoria, during the Crimean War, 1854

A beaten general is disgraced forever.

> Ferdinand Foch, *Precepts*, 1919

The winner is asked no questions – the loser has to answer for everything.

> Sir Ian Hamilton, *Gallipoli Diary*, 1920

Errors and defeats are more obviously illustrative of principles than successes are ... Defeat cries aloud for explanation; whereas success, like charity, covers a multitude of sins.

> Alfred Thayer Mahan, *Naval Strategy*, 1911

France has lost a battle. But France has not lost the war.
> President Charles de Gaulle, broadcast to the French
> People, 18 June 1940

Man in war is not beaten, and cannot be beaten, until he owns himself beaten.
> B.H. Liddell Hart, *Thoughts on War*, 1944

Fearful are the convulsions of defeat.
> Sir Winston Churchill, *The Gathering Storm*, 1948

Defence

Petty geniuses attempt to hold everything; wise men hold fast to the key points. They parry great blows and scorn little accidents. There is an ancient apothegm: he who would preserve everything, preserves nothing. Therefore, always sacrifice the bagatelle and pursue the essential.
> Frederick the Great, *Instructions for His Generals*, 1747

Defence is a shield made up of well directed blows.
> Karl von Clausewitz, *On War*, 1832

A swift and vigorous transition to attack – the flashing sword of vengeance – is the most brilliant point of the defensive.
> Karl von Clausewitz, *On War*, 1832

A clever military leader will succeed in many cases in choosing defensive positions of such an offensive nature from the strategic point of view that the enemy is compelled to attack us in them.
> Helmuth von Moltke ('The Elder'), 1800–91

In war, the defensive exists mainly that the offensive may act more freely.
> Alfred Thayer Mahan, *Naval Strategy*, 1911

Every position must be held to the last man; there must be no retirement. With our backs to the wall, and believing in the justice of our cause, each one of us must fight on to the end.
> Sir Douglas Haig, order to the British Army in France,
> 12 April 1918

Distances should be reduced. Reserves should be held closer in hand to prevent the enemy breaking through and to check quickly his initial penetrations.

B.H. Liddell Hart, *Thoughts on War*, August 1920

In deciding whether to occupy commanding positions in enclosed country, their increased field of fire and observation must be weighted with their disadvantages as an easy objective and artillery target for the enemy.

B.H. Liddell Hart, *Thoughts on War*, August 1920

Every development or improvement in firearms favours the defensive.

Giulio Douhet, *The Command of the Air*, 1921

In planning defensive positions one should ensure that they do more than cover the enemy's natural route of advance – remembering that a real tactician will seek to avoid the line of expectation.

B.H. Liddell Hart, *Thoughts on War*, August 1928

Obstruction is the natural antidote to the power of delivering mobile strokes which mechanization has revived ... the defender's mechanized troops, having less risk of being checked by hostile obstructions, may be switched to meet the danger faster than the mechanized attackers can develop it.

B.H. Liddell Hart, *Thoughts on War*, July 1936

We shall defend every village, every town and every city. The vast mass of London itself, fought street by street, could easily devour an entire hostile army; and we would rather see London laid in ruins and ashes than that it should be tamely and abjectly enslaved.

Sir Winston Churchill, broadcast to the British people, 14 July 1940

Defiance

There is plenty of time to win this game, and to thrash the Spaniards too.

Attributed to Sir Francis Drake, Plymouth Hoe, 1588

Ruthven Redoubt, 30 August 1745

Hon General – This goes to acquaint you that yesterday there appeared in the little town of Ruthven about three hundred of the enemy, and sent proposals to me to surrender the redoubt upon Condition that I should have liberty to carry off bags and baggage. My answer was, 'I am too old a soldier to surrender a garrison of such strength without bloody noses!' They threatened to hang me and my men for refusal. I told them I would take my chance. This morning they attacked me about twelve o'clock with about one hundred and fifty men; they attacked the fore-gate and sally-port. They drew off about half an hour after three. I expect another visit this night, but I shall give them the warmest reception my weak party can afford. I shall hold out as long as possible.

I conclude, Honourable General, with great respects,
Your most humble servant,
J. Molley, Sergt. 6th [Foot]
Quoted in Thomas Gilby, *Britain at Arms*, 1953

Nuts!

Major General Anthony McAuliffe, US Army. This was his reply to a German demand that he surrender his encircled force at Bastogne, 23 December 1944

Determination

In battle, two moral forces, even more than two material forces, are in conflict. The stronger conquers. The victor has often lost ... more men than the vanquished ... With equal or even inferior power of destruction, he will win who is determined to advance.

Ardant du Picq, 1821–70, *Battle Studies*

It is not the actual military structure of the moment that matters but rather the will and determination to use whatever military strength is available.

Adolf Hitler, *Mein Kampf*, 1925

I shall return.

Douglas McArthur, on departure from Corregidor, 11 March 1942

Difficulties

I am not come forth to find difficulties, but to remove them.
> Horatio Nelson, 1758–1805

I wish I could persuade you to try to overcome the difficulties instead of merely entrenching yourself behind them.
> Winston Churchill, memorandum for Minister of
> Agriculture, 28 February 1943

Disarmament

He maketh wars to cease in all the world:
he breaketh the bow, and snappeth the spear in sunder, and
burneth the chariots in the fire.
> Psalm 46

From time immemorial, the idea of disarmament has been one of the most favoured forms of diplomatic dissimulation of the true motives and plans of those governments which have been seized by sudden 'love of peace'. This phenomenon is very understandable. Any proposal for the reduction of armaments could invariably count upon broad popularity and support from public opinion.
> E.V. Tarle, 1862–1917, *History of Diplomacy*

I am quite prepared to disarm – provided that the others disarm first.
> Benito Mussolini, 1883–1945

Discipline

If troops are punished before their loyalty is secured they will be disobedient. If not obedient, it is difficult to employ them. If troops are loyal, but punishments are not enforced, you cannot employ them. Thus, command them with civility and imbue them uniformly with martial ardour and it may be said that victory is certain ... When orders are consistently trustworthy and observed, the relationship of a commander with his troops is satisfactory.

Sun Tzu, 400–320 BC, *The Art of War*, ix

No state can be either happy or secure that is remiss and negligent in the discipline of its troops.

Vegetius, *The Military Institutions of the Romans*, i, AD 378

As for Colonel Cromwell, he hath 2,000 brave men, well disciplined; no man swears but he pays his twelvepence; if he be drunk, he is set in the stocks, or worse; if one calls the other roundhead he is cashiered ... How happy were it if all the forces were thus disciplined.

Puritan newsletter, May 1643

The Romans conquered all peoples by their discipline. In the measure that it became corrupted their success decreased. When the Emperor Gratian permitted the legions to give up their cuirasses and helmets because the soldiers complained that they were too heavy, all was lost. The barbarians whom they had defeated for so many centuries vanquished them in their turn.

Maurice de Saxe, *Reveries*, 1732

After the organization of troops, military discipline is the first matter that presents itself. It is the soul of armies. If it is not established with wisdom and maintained with unshakeable resolution you will have no soldiers. Regiments and armies will be only contemptible, armed mobs, more dangerous to their own country than to the enemy ... It has always been noted that it is with those armies in which the severest discipline is enforced that the greatest deeds are performed.

Maurice de Saxe, *Reveries*, 1732

Discipline is simply the art of inspiring more fear in the soldiers of their officers than of the enemy.

Helvetius, *On the Spirit*, 1758

Men accustomed to unbounded freedom, and no control, cannot brook the Restraint which is indispensably necessary to the good order and government of an Army.

George Washington, letter to the President of Congress, 1776

Discipline begins in the Wardroom. I dread not the seamen. It is the indiscreet conversations of the officers and their presumptuous discussions of the orders they receive that produce all our ills.

> Lord St Vincent, 1735–1823

Discipline is summed up in one word, obedience.

> Lord St Vincent, 1735–1823

The officers of companies must attend to the men in their quarters as well as on the march, or the army will soon be no better than a banditti.

> The Duke of Wellington, General Order of 19 May 1800

Popularity, however desirable it may be to individuals, will not form, or feed, or pay an army; will not enable it to march, and fight; will not keep it in a state of efficiency for long and arduous service.

> The Duke of Wellington, letter from Portugal, 8 April 1811

We have in the service the scum of the earth as common soldiers; and of late years we have been doing everything in our power, both by law and by publications, to relax the discipline by which alone such men can be kept in order. The officers of the lower ranks will not perform the duty required from them for the purpose of keeping their soldiers in order; and it is next to impossible to punish any officer for neglects of this description. As to the non-commissioned officers, as I have repeatedly stated, they are as bad as the men, and too near them in point of pay and situation, by the regulations of late years, for us to expect them to do anything to keep the men in order.

> The Duke of Wellington, after Vitoria, 21 June 1813

My son, put a little order into your corps, it wants it badly. The Italians in particular commit atrocities, robbing and pillaging wherever they go. Shoot a few of them. Your affectionate father.

> Napoleon I to Eugene, May 1813

Discipline is not made to order, cannot be created offhand; it is a matter of the institution of tradition. The Commander must have absolute confidence in his right to command, must have the habit of command, pride in commanding.

Ardant du Picq, 1821–70, *Battle Studies*

Discipline and blind obedience are things which can be produced and given permanence only by long familiarity.

Wilhelm I of Prussia, 1797–1888

To be disciplined ... means that one frankly adopts the thoughts and views of the superior in command, and that one uses all humanly practicable means in order to give him satisfaction.

Ferdinand Foch, *Precepts*, 1919

We was rotten 'fore we started – we was never disciplined;
We made it out a favour if an order was obeyed.
Yes, every little drummer 'ad 'is rights an' wrongs to mind,
So we had to pay for teachin' – an' we paid!

Rudyard Kipling, *That Day*, 1895

If you can't get them to salute when they should salute and wear the clothes you tell them to wear, how are you going to get them to die for their country?

Lieutenant-General George Patton, 1885–1945

Discipline is teaching which makes a man do something which he would not, unless he had learnt that it was the right, the proper, and the expedient thing to do. At its best, it is instilled and maintained by pride in oneself, in one's unit, in one's profession; only at its worst by a fear of punishment.

Sir Archibald Wavell, *Soldiers and Soldiering*, 1953

To obey God's orders as delivered by conscience – that is duty; to obey man's orders as issued by rightful authority – that is discipline. The foundation of both alike is denial of self for a higher good. Unless the lesson of duty be first well learned, the lesson of discipline can be put imperfectly understood.

Sir John Fortescue, 1859–1933, *A Gallant Company*

Divide

I can only advise the party on the defensive not to divide his forces too much by attempting to cover every point.
Antoine Henri Jomini, *Summary of the Art of War*, 1838

Divide and rule [sometimes rendered as 'Divide and conquer'].
Ancient maxim

Division

The Pope! How many divisions has he got?
J.V. Stalin, to Pierre Laval during conversations in Moscow, May 1935, quoted in Sir Winston Churchill, *The Gathering Storm*, 1948

… the smallest formation that is a complete orchestra of war and the largest in which every man can know you.
Sir William Slim, *Defeat into Victory*, 1956

Doctrine

A doctrine of war consists first in a common way of objectively approaching the subject; second, in a common way of handling it.
Ferdinand Foch, *Precepts*, 1919

Official manuals, by the nature of their compilation, are merely registers of prevailing practice, not the log-books of a scientific study of war.
B.H. Liddell Hart, *Thoughts on War*, 1944

Generals and admirals stress the central importance of 'doctrine'. Military doctrine is the 'logic of their professional behaviour. As such, it is a synthesis of scientific knowledge and expertise on the one hand, and of traditions and political assumptions on the other.
Morris Janowitz, *The Professional Soldier*, 1960

Doctrine is indispensable to an army … Doctrine provides a military organization with a common philosophy, a common language, a common purpose, and a unit of effort.

> General George Decker, US Army Address, Command and General Staff College, Fort Leavenworth, Kansas, 16 December 1960

Dress

The better you dress a soldier, the more highly will he be thought of by the women, and consequently by himself.

> Field Marshal Lord Wolseley, *The Soldier's Pocketbook*, quoted in Thomas Gilby, *Britain at Arms*, 1953

Drill

Drill is necessary to make the soldier steady and skilful, although it does not warrant exclusive attention.

> Maurice de Saxe, *Reveries*, 1732

The exterior splendour, the regularity of movements, the adroitness and at the same time firmness of the mass – all this gives the individual soldier the safe and calming conviction that nothing can withstand his particular regiment or battalion.

> Colmar von der Goltz, 1843–1916, *Rossbach and Jena*

Drum

Not a drum was heard, not a funeral note,
As his corse to the rampart we hurried.

> Charles Wolfe, 1791–1823, *The Burial of Sir John Moore*

Oh, it's Tommy this an' Tommy that an'
'Tommy, go away',

But it's 'Thank you Mister Atkins' when the band begins to
  play.
Then it's Tommy this an' Tommy that, an'
'Tommy 'ows yer soul?'
But it's 'Thin red line of 'eroes' when the drums begins to roll
                        Rudyard Kipling, 1865–1936, *Tommy*

Drunkenness

The British soldiers are fellows who have all enlisted for drink –
that is the plain fact – they have all enlisted for drink.
                    The Duke of Wellington, letter from Portugal, 1811

'Drunk and resisting the Guard!'
Mad drunk and resisting the Guard –
'Strewth, but I socked it them hard!
So it's pack-drill for me and a fortnight's CB
For 'Drunk and resisting the Guard.'
                        Rudyard Kipling, *Cells*, 1892

Duty

In doing what we ought, we deserve no praise, because it is our
duty.
                        St Augustine, AD 354–430

Duty is the great business of a sea-officer; all private
considerations must give way to it, however painful it may be.
                    Horatio Nelson, letter to Frances Nisbet, 1786

England expects that every man will do his duty.
                        Horatio Nelson, to his flag-lieutenant, aboard HMS
                        Victory, before Trafalgar, 21 October 1805

Thank God I have done my duty.
                        Horatio Nelson, while lying mortally wounded in the
                        cockpit, HMS *Victory*, Trafalgar, 21 October 1805

Whatever happens, you and I will do our duty.

> The Duke of Wellington, remark to Lord Uxbridge on the eve of Waterloo, 17 June 1815

We gave thanks to God for the noblest of all His blessings, the sense that we had done our duty.

> Sir Winston Churchill, 1874–1965

If I do my full duty, the rest will take care of itself.

> Lieutenant General George Patton, diary entry before the North African landings, 8 November 1942

I pray daily to do my duty, retain my self-confidence and accomplish my destiny.

> Lieutenant General George Patton, diary entry, 20 June 1943

Duty is never simple, never easy, and rarely obvious.

> Jean Dutourd, *Taxis of the Marne*, 1957

Economy of Effort

I could lick those fellows any day, but it would cost me 10,000 men, and, as this is the last army England has, we must take care of it.

> The Duke of Wellington, of Massena's army in
> Portugal, October 1810

A war should only be undertaken with forces proportioned to the obstacles to be overcome.

> Napoleon 1, *Maxims of War*, 1831

The principle of economy of force is the art of pouring out all one's resources at a given moment on one spot.

> Ferdinand Foch, *Precepts*, 1919

... the art of making the weight of all one's forces successively bear on the resistances one may meet.

> Ferdinand Foch, *Principles of War*, 1920

To me, an unnecessary action, or shot, or casualty, was not only waste but sin.

> T.E. Lawrence, 1888–1935, *Seven Pillars of Wisdom*

We used the smallest force, in the quickest time, in the farthest place.

> T.E. Lawrence, 1888–1935, *Seven Pillars of Wisdom*

Elite Forces

It is better to have a small number of well-kept and well disciplined troops than to have a great number who are neglected in these matters. It is not big armies that win battles; it is the good ones.

Maurice de Saxe, *Reveries*, 1732

We are the pilgrims, masters. We shall go always a little further. It may be beyond that last blue mountain, bar'ed with snow, across that angry or that glimmer sea.

James Elroy Flecker, 1884–1915, *Hassan*

Enemy

A general in all of his projects should not think so much about what he wishes to do as what his enemy will do; that he should never underestimate this enemy, but he should put himself in his place to appreciate difficulties and hindrances the enemy could interpose; that his plans will be deranged at the slightest event if he has not foreseen everything and if he has not devised means with which to surmount the obstacles.

Frederick the Great, *Instructions for His Generals*, 1747

You must not fight too often with one enemy, or you will teach him all your art of war.

Napoleon I, 1769–1821, quoted in Dictionary of
Military and Naval Quotations

However absorbed a commander may be in the elaboration of his own thoughts, it is sometimes necessary to take the enemy into account.

Attributed to Sir Winston Churchill, 1874–1965

I drink to the memory of a gallant and honourable foe.

Graf von Spee at Valparaiso when invited to drink to
the damnation of the Royal Navy by a German
resident after the battle of Coronel. Quoted in papers
belonging to Vice-Admiral KGB Dewar.

Experienced military men are familiar with the tendency that always has to be watched in staff work, to see all our own difficulties but to credit the enemy with the ability to do things we should not dream of attempting.

Sir John Slessor, *Strategy for the West*, 1954

Engineer

The rough ways shall be made smooth.
Luke 3

Which of you, intending to build a tower, sitteth not down first, and countest the cost, whether he have sufficient to finish it?
Luke 14

For 'tis the sport to have the engineer
Hois'd with his own petard.
William Shakespeare, *Hamlet*, 1600

In the final analysis no military force gets far without its engineers, who were the first among educated soldiers and remain to this day among the elite of every well-founded army.
Anon.

The three great sapper lies: my best friend is a gunner, my cheque is in the post, and I'll get onto it straight away.
Anon.

Envelopment

The deep envelopment based on surprise, which severs the enemy's supply lines, is and always has been the most decisive manoeuvre of war. A short envelopment, which fails to envelop and leaves the enemy's supply system intact, merely divides your own forces and can lead to heavy loss and even jeopardy.

General Douglas MacArthur, conference before Inchon, 23 August 1950

Equipment

There are five things the soldier should never be without – his musket, his ammunition, his rations (for at least four days), and his entrenching tool. The knapsack may be reduced to the smallest size possible ... but the soldier should always have it with him.

Napoleon I, *Maxims of War*, 1831

Esprit de Corps

That silly, sanguine notion, which is firmly entertained here, that one Englishman can beat three Frenchmen, encourages, and has sometimes enabled, one Englishman, in reality, to beat two.

Lord Chesterfield, *Letters*, 7 February 1749

All that can be done with the soldier is to give him esprit de corps – i.e., a higher opinion of his own regiment than all the other troops in the country.

Frederick the Great, *Military Testament*, 1768

Now is the time to popularize with the troops by giving to all regiments and units the little badges and distinctions they like so much ... All regimental distinctions should be encouraged.

Sir Winston Churchill, to Secretary of State for War, July 1940

Living in an atmosphere of soldierly duty and esprit de corps permeates the soul, where drill merely attunes the muscles.

B.H. Liddell Hart, *Thoughts on War*, 1944

Evacuations

Wars are not won by evacuations.

Sir Winston Churchill, to the House of Commons on Dunkirk, June 1940

Example

A brave captain is as a root, out of which, as branches, the courage of his soldiers doth spring.

Sir Philip Sidney, 1554–86, *Dictionary of Naval and Military Quotations*

No man should possess him with any appearance of fear, lest he, by shewing it, should dishearten his army.

William Shakespeare, *King Henry V, 1598*

If you choose godly, honest men to be captains of Horse, honest men will follow them.

Oliver Cromwell, to Sir William Springe, September 1643

As I would deserve and keep the kindness of this army, I must let them see that when I expose them, I would not exempt myself.

Duke of Marlborough, 1650–1722, *Dictionary of Naval and Military Quotations*

Let your character be above reproach, for that is the way to earn men's obedience.

Mathias von Schulenburg, to Marshal Saxe when a young officer, 1709

I am the very model of a modern major-general.

W.S. Gilbert, *The Pirates of Penzance*, 1880

With two thousand years of examples behind us, we have no excuse when fighting, for not fighting well.

T.E. Lawrence, 1888–1935, Staff College papers

Officers should live under the same conditions as their men, for that is the only way in which they can gain from their men the admiration and confidence so vital in war. It is incorrect to hold a theory of equality in all things, but there must be equality of existence in accepting the hardships and dangers of war.

Mao Tse-tung, *On Guerrilla Warfare*, 1937

Be an example to your men, both in your duty and in private life. Never spare yourself, and let the troops see that you don't in your endurance of fatigue and privation. Always be tactful and well mannered and teach your subordinates to be the same. Avoid excessive sharpness or harshness of voice, which usually indicates the man who has shortcomings of his own to hide.

> Field Marshall Erwin Rommel, remarks to cadets
> passing out of Wiener Neustadt Military School, 1938

In moments of panic, fatigue or disorganization, or when something out of the ordinary has to be demanded from [his troops], the personal example of the commander works wonders, especially if he has had the wit to create some sort of legend round himself.

> Field Marshal Erwin Rommel, *The Rommel Papers*,
> 1953

Exhortation

You are ordered abroad as a soldier of the King to help your French comrades against the invasion of a common enemy. You have to perform a task which will need your courage, your energy, your patience. Remember that the honour of the British Army depends on your individual conduct. It will be your duty not only to set an example of discipline and perfect steadiness under fire but also to maintain the most friendly relations with those whom you are helping in his struggle ... Do your duty bravely. Fear God. Honour the King.

> Lord Kitchener, to the soldiers of the British
> Expeditionary Force, on embarking for France,
> August 1914

Experience

Military men who spend their lives in the uniform of their country acquire experience in preparing for war and waging it.

No theoretical studies, no intellectual attainments on the part of the layman can be a substitute for the experience of having lived and delivered under the stress of war.

General Maxwell Taylor, graduation address at West Point, June 1963

Families

He that hath wife and children hath given hostage to fortune, for they are impediments to great enterprises, either of virtue or of mischief.

> Francis Bacon, 1561–1626, *Essays*

Let soldiers marry; they will no longer desert. Bound to their families, they are bound to their country.

> François-Marie Arouet Voltaire, 1694–1778, *Satirical Dictionary*

A wife is a useless piece of furniture for a soldier.

> Madame de Pompadour, 1721–64, letter to the Duchess of d'Estrées

If any of the soldiers' wives or children happen to be taken ill, never give them any assistance. You receive no pence from them, and you know *ex nihilo nihil sit.*

> Francis Grose, *Advice to the Officers of the British Army*, 1782

The duty of the captain of a battleship in wartime is incessant, requiring the most constant and unremitting attention, and the best first lieutenant in the world cannot be sufficient substitute for him, in that his whole body and soul should be in it which, with his wife and family near him, never has been, or can be; and, unless the husband and wife can make up their minds to separation during war, it would be unwise and unjust to the Service, and to me, for him to delay one day his intention of retiring.

> Lord St Vincent, letter to a captain's wife, 1800

Marriage is good for nothing in the military profession.

> Napoleon I, *Political Aphorisms*, 1848

Fear

The man that is roused neither by glory nor danger it is vain to exhort; terror closes the ears of the mind.

> Sallust, *The Catiline Conspiracy*, 43 BC

Nothing is to be feared but fear.

> Francis Bacon, 1521–66, Essays

To fear the foe, since fear oppresseth strength,
Gives in your weakness strength unto your foe.
And so your follies fight unto yourself.
Fear and be slain; no worse can come to fight:
And fight and die is death destroying death;
Where fearing dying pays death servile breath.

> William Shakespeare, *King Richard II*, 1595

Nothing so much to be feared as fear.

> Henry David Thoreau, 1817–62, *Journal*

The first duty of a man is still that of subduing fear.

> Thomas Carlyle, 'Heroes and Hero Worship', lecture, 1840

All men are frightened. The more intelligent they are, the more they are frightened. The courageous man is the man who forces himself, in spite of his fear, to carry on. Discipline, pride, self-respect, self-confidence, and love of glory are attributes which will make a man courageous even when he is afraid.

> Lieutenant General George Patton, *War As I Knew It*, 1947

No sane man is unafraid in battle.

> Lieutenant-General George Patton, *War As I Knew It*, 1947

Fifth Column

We have four columns advancing upon Madrid. The fifth column will rise at the proper time.

> Emilio Mola, radio broadcast, October 1936 (Ernest Hemingway popularized the phrase in the title of his 1938 play, *The Fifth Column*)

When I wage war ... in the mist of peace troops will suddenly appear, let us say, in Paris. They will wear French uniforms. They will march through the streets in broad daylight. No one will stop them. Everything has been thought out, prepared to the last detail. They will march to the headquarters of the General Staff. They will occupy the ministries, the Chamber of Deputies ... Peace will be negotiated before the war has begun. We shall have enough volunteers, men like our SA, trustworthy and ready for any sacrifice. We shall send them across the border in peacetime. Gradually the place of artillery preparation for frontal attack by the infantry will in future be taken by revolutionary propaganda, to break down the enemy psychologically before the armies begin to function at all.

Adolf Hitler, 1889–1945, *Mein Kampf*

Fight, Fighting

Fight on, my men, Sir Andrew says,
A little I'm hurt, but not yet slain;
I'll lie me down and bleed awhile,
And then I'll rise and fight again.

Ballad, *Sir Andrew Barton, c.*1550

Fight to the last gasp.

William Shakespeare, *I King Henry VI*, 1591

Fight, gentlemen of England! Fight, bold yeomen!
Draw, archers, your arrows to the head!
Spur your proud horses hard, and ride in blood;
Amaze the welkin with your broken staves!

William Shakespeare, *King Richard II*, 1592

I'll fight, till from my bones my flesh be hacked.

William Shakespeare, *Macbeth*, 1605

I have not yet begun to fight.

John Paul Jones, in reply to the hail, 'Have you struck?' from Captain Richard Pearson, RN, commanding HMS *Serapis*, off Flamborough Head, 23 September 1775

War in its literal meaning is fighting, for fighting alone is the efficient principle in the manifold activity which in a wide sense is called war.
> Karl von Clausewitz, *On War*, 1832

Fight the good fight with all thy might.
> J.S.B. Monsell, *Fight the Good Fight*, 1863

If we must be enemies, let us be men, and fight it out as we propose to do, and not deal in hypocritical appeals to God and humanity.
> W.T. Sherman, letter to General John Hood, Atlanta, 10 September 1864

You will see me in the field fighting for your independence long after you and your party who make war with your mouths have fled the country.
> Jacobus de la Rey to President Kruger, eve of the Boer War, 1899

I would fight without a break. I would fight in front of Amiens. I would fight in Amiens. I would fight behind Amiens. I would never surrender.
> Ferdinand Foch, to Sir Douglas Haig, 26 March 1918

I shall fight before Paris, I shall fight in Paris, I shall fight behind Paris. (Je me bats devant Paris, je me bats à Paris, je me bats derrière Paris.)
> Georges Clemençeau, to the Chamber of Deputies, 4 June 1918 (cf. Foch)

That this house will in no circumstances fight for its King and country.
> Motion passed at the Oxford Union, 9 February 1933

If it is thought best for France in her agony that her Army should capitulate let there be no hesitation on our account, because whatever you may do, we shall fight on for ever and ever and ever.
> Sir Winston Churchill, message to Paul Reynaud, June 1940

We shall go on to the end, we shall fight in France, we shall fight on the seas and oceans, we shall fight with growing confidence and growing strength in the air, we shall defend our island, whatever the cost may be, we shall fight on the beaches, we shall fight on the landing grounds, we shall fight in the fields and in the streets, we shall fight in the hills; we shall never surrender.

> Sir Winston Churchill, broadcast, 4 June 1940, after Dunkirk

When I warned them [the French Government] that Britain would fight on alone whatever they did, their Generals told their Prime Minister and his divided Cabinet: 'In three weeks England will have her neck wrung like a chicken.' Some chicken, some neck!

> Sir Winston Churchill, to the Canadian Parliament, 30 December 1941

We will fight the enemy where we now stand; there will be no withdrawal and no surrender.

> Montgomery of Alamein, to the Eighth Army after assuming command, 13 August 1942

Still, if you will not fight for the right when you can easily win without bloodshed; if you will not fight when your victory will be sure and not too costly; you may come to the moment when you will have to fight with all the odds against you and only a precarious chance of survival. There may even be a worse case. You may have to fight when there is no hope of victory, because it is better to perish than live as slaves.

> Sir Winston Churchill, *The Gathering Storm*, 1948

Fire, Firepower

Fire! And may God have mercy on their guilty souls!

> General Pendleton, Captain of the Rockridge Artillery at Bull Run, 21 July 1861

Silent till you see the whites of their eyes.

> Prince Charles of Prussia, at Jägerndorf, 23 May 1745

Don't fire 'til you see the whites of their eyes.

> William Prescott, order at Bunker Hill, 17 June 1775

It is firepower, and firepower that arrives at the right time and place, that counts in modern war.

B.H. Liddell Hart, *Thoughts on War*, 1944

The tendency towards under-rating fire-power ... has marked every peace interval in modern military history.

B.H. Liddell Hart, *Thoughts on War*, 1944

The greater the visible effect of fire on the attacking infantry, the firmer grows the defenders' morale, whilst a conviction of the impregnability of the defence rapidly intensifies in the mind of the attacking infantryman.

B.H. Liddell Hart, *Thoughts on War*, 1944

Fleet

Most men were in fear that the French would invade; but I was always of another opinion, for I always said that whilst we had a fleet in being, they would not dare to make an attempt.

Admiral Sir Arthur Torrington, RN, in a report, 1690, justifying his decision not to engage a superior French fleet, probably the origin of the Fleet in Being concept

A fleet of British ships of war are the best negotiators in Europe.

Horatio Nelson, letter to Lady Hamilton, March 1801

The moral effect of an omnipresent fleet is very great, but it cannot be weighed against a main fleet known to be ready to strike and able to strike hard.

Sir John Fisher, to Lord Stamfordham, 25 June 1912

Flexibility

No plan of operations can look with any certainty beyond the first meeting with the major forces of the enemy. The commander is compelled ... to reach decisions on the basis of situations which cannot be predicted.

Helmuth von Moltke, 1800–91, letter to a friend

Force

The use of force is but temporary. It may subdue for a moment; but it does not remove the necessity of subduing again; and a nation is not governed, which is perpetually to be conquered.
Edmund Burke, 22 March 1775

The maximum use of force is in no way incompatible with the simultaneous use of the intellect.
Karl von Clausewitz, *On War*, 1832

Forethought

A C-in-C must draw up a master plan for the campaign he envisages and he must always think and plan two battles ahead – the one he is preparing to fight and the next one – so that successes gained in one battle may be used as a springboard for the next.
Montgomery of Alamein, *Memoirs*, 1958

Fortifications

Field fortification, when well conceived, is always useful and never harmful.
Napoleon I, *Maxims of War*, 1831

The principles of field fortification need to be perfected. This part of the art of war is susceptible of making great progress.
Napoleon I, at St Helena, 1816

It is wise to regard fortification as only a means to advantageous battle – a sponge to absorb the enemy's force with small diversion of one's own.
B.H. Liddell Hart, *Thoughts on War*, 1933

Friction

The friction inherent in the tremendous war-machine of an armed power is so great in itself that it should not be increased unnecessarily.

> Karl von Clausewitz, *Principles of War*, 1812

Everything is very simple in war, but the simplest thing is difficult. These difficulties accumulate and produce a friction which no man can imagine exactly who has not seen war.

> Karl von Clausewitz, *On War*, 1832

The military machine is composed of individuals, everyone of which retains his potential for friction.

> Karl von Clausewitz, *On War*, 1832

Front

The army report confined itself to a single sentence: All quiet on the Western Front.

> Erich Maria Remarque, *Nothing New in the West* in *Westen Nichts Neues*, 1929

Look to your front.

> Drill Command of the Rifle Brigade, British Army

Gallantry

Nothing in his life
Became him like the leaving of it; he died
As one that had studied in his death
To throw away the dearest thing he owed
As 'twere a careless trifle.
> William Shakespeare, *Macbeth*, 1605

Nothing on earth is so stupid as a gallant officer.
> The Duke of Wellington, after the battle of Fuentes de
> Oñoro, 3 May 1811

Hereabouts died a very gallant gentleman, Captain L.E.G. Oates of the Inniskilling Dragoons. In March 1912, returning from the Pole, he walked willingly to his death in a blizzard, to try and save his comrades, beset by hardships.
> Surgeon Captain E.L. Atkinson 1882–1929 and Apsley
> Cherry-Garrard 1882–1959, epitaph on a cairn and
> cross erected in the Antarctic, November 1912

General Staff

The ideal General Staff should, in peace time, do nothing! They deal in an intangible stuff called thought. Their main business consists in thinking out what an enemy may do and what their Commanding Generals ought to do, and the less they clank their spurs the better.
> Sir Ian Hamilton, *The Soul and Body of an Army*, 1921

Doesn't it make you surer you were right, to see all the General Staff opposing you?
> T.E. Lawrence, letter to Ernest Thurtle, 2 May 1930

The General Staff was truly ... an all-powerful military priesthood, linked by ties of intellectual and professional comradeship. A corps of directors, a society within a society, they were to the German Army what the Jesuits at their political zenith were to the Church of Rome.

B.H. Liddell Hart, *Thoughts on War*, 1944

Generals

The General must know how to get his men their rations and every other kind of stores needed for war. He must have imagination to originate plans, and practical sense and energy to carry them through. He must be observant, untiring, shrewd; kindly and cruel; simply and crafty; a watchman and a robber; lavish and miserly; generous and stingy; rash and conservative. All these and many other qualities, natural and acquired, he must have. He should also, as a matter of course, know his tactics; for a disorderly mob is no more an army than a heap of building materials is a house.

Socrates, 469–399 BC

The general who in advancing does not seek personal fame, and in withdrawing is not concerned with avoiding punishment, but whose only purpose is to protect the people and promote the best interests of his sovereign, is the precious jewel of the state. Because such a general regards his men as infants they will march with him into the deepest valleys. He treats them as his own beloved sons and they will die with him.

Sun Tzu, 400–320 BC, *The Art of War*, x

A general must be continent, sober, frugal , hard-working, middle-aged, eloquent, father of a family, and member of an illustrious house. Soldiers do not like being under the command of one who is not of good birth. In addition, a general should be polite, affable, easy of approach, and cool-headed.

Onasander, first century AD

114

I have formed a picture of a general commanding, which is not chimerical. I have seen such men. The first of all qualities is Courage. Without this the others are of little value, since they cannot be used. The second is Intelligence, which must be strong and fertile in expedients. The third is Health.

> Maurice de Saxe, *Reveries*, 1732

When I reflect upon the characters and attainments of some of the general officers of this army, and consider that these are the persons on whom I am to rely to lead columns against the French, I tremble; and as Lord Chesterfield said of the generals of his day, 'I only hope that when the enemy reads the list of their names, he trembles as I do'.

> The Duke of Wellington in a dispatch to Torrens, 29 August 1810. Usually quoted as 'I don't know what effect these men will have upon the enemy, but, by God, they frighten me.' Also attributed to George III.

Many of the greatest military commanders owe their exaltation and celebrity to the Art of letter writing.

> Colonel J.G.D. Tucker, *Advice to Young Officers*, 1826

It is exceptional and difficult to find in one man all the qualities necessary for a great general. What is most desirable, and which instantly sets a man apart, is that his intelligence or talent, are balanced by his character of courage.

> Napoleon I, *Maxims of War*, 1831

Generals have never risen from the very learned or really erudite class of officers, but have been mostly men who, from the circumstances of their position, could not have attained to any great amount of knowledge.

> Karl von Clausewitz, *On War*, 1832

People are rather inclined to look upon a subordinate general grown grey in the service, and in whom constant discharge of routine duties had produced a decided poverty of mind, as a man of failing intellect, and, with all respect for his bravery, to laugh at his simplicity.

> Karl von Clausewitz, *On War*, 1832

The most essential qualities of a general will always be: first, a high moral courage, capable of great resolution; second, a physical courage which takes no account of danger. His scientific or military acquirements are secondary to these.

> Antoine Henri Jomini, *Summary of the Art of War*, 1838

Let no man be so rash as to suppose that, in donning a general's uniform, he is forth-with competent to perform a general's functions.

> Dennis Hart Mahan, 1802–71 (Professor Mahan was the father of Alfred Thayer Mahan, the naval strategist)

I am the very model of a modern major-general.
I've information vegetable, animal and mineral.
I know the kings of England, and I quote the fights historical
From Marathon to Waterloo, in order categorical.

> W.S. Gilbert, *The Pirates of Penzance*, 1880

'Good morning! Good morning!' the General said
When we met him last week on the way to the Line.
Now the soldiers he smiled at are most of 'em dead
And we're cursing his staff for incompetent swine.
'He's a cheery old card', grunted Harry to Jack,
As they slogged up to Arras with rifle and pack,
But he did for them both with his plan of attack.

> Siegfried Sassoon, *The General*, 1917

Battles are lost or won by generals, not by the rank and file.

> Ferdinand Foch, *Principles of War*, 1920

It takes 15,000 casualties to train a major general.

> Attributed to Ferdinand Foch, 1851–1929

Generalship

He who knows when he can fight and when he cannot will be victorious.

> Sun Tzu, 400–320 BC, *The Art of War*, iii

When campaigning, be swift as the wind; in leisurely march, majestic as the forest; in raiding and plundering, like fire; in standing, firm as the mountains. As unfathomable as the clouds, move like a thunderbolt.

Sun Tzu, 400–320 BC, *The Art of War*, vii

The object of a good general is not to fight, but to win. He has fought enough if he gains a victory.

Duke of Alba, *c.*1560

The coup d'oeuil is a gift of God and cannot be acquired; but if professional knowledge does not perfect it, one only sees things imperfectly and in a fog, which is not enough in these matters where it is so important to have a clear eye ... To look over a battlefield, to take in at the first instance the advantages and disadvantages is the great quality of a general.

Chevalier Folard, *New Discoveries about War*, 1724

The first quality of a general in chief is a great knowledge of the art of war. This is not intuitive, but the result of experience. A man is not born a commander. He must become one. Not to be anxious; to be always cool; to avoid confusion in his commands; never to change countenance; to give his orders in the midst of battle with as much composure as if he were perfectly at ease. These are the proofs of valour in a general.

Count de Montecucculi, *War Commentaries*, 1740

We should try to make war without leaving anything to chance. In this lies the talent of a general.

Maurice de Saxe, *Reveries*, 1732

When the Duke of Cumberland has weakened his army sufficiently, I shall teach him that a general's first duty is to provide for its welfare.

Maurice de Saxe, before Laufeldt, 1747

A great captain can only be formed by long experience and intense study.

Archduke Charles of Austria, 1771–1841

The real reason why I succeeded in my own campaigns is because I was always on the spot.

The Duke of Wellington, letter, 1805

I engage and after that I see what to do. (Je m'engage, et après ca, je voie).

> Napoleon I, remark during the Italian campaign, 1796

The only true wisdom in a general is determined courage.

> Napoleon I, *Maxims of War*, 1831

The first qualification in a general is a cool head – that is, a head which receives accurate impressions, and estimates things and objects at their real value. He must not allow himself to be elated by good news, or depressed by bad.

> Napoleon I, *Maxims of War*, 1831

The art of war is simple enough. Find out where your enemy is. Get at him as soon as you can. Strike at him at hard as you can and keep moving on.

> General Ulysses Grant, 1822–85

The modern commander-in-chief is no Napoleon who stands with his brilliant suite upon a hill ... The commander is farther to the rear in a house with roomy offices, where telegraph and wireless, telephone and signalling instruments are at hand, while a fleet of automobiles and motorcycles, ready for the longest trips, wait for orders. Here, in a comfortable chair before a large table, the modern Alexander overlooks the entire battlefield on a map. From here he telephones inspiring words, and here he receives the reports from army and corps commanders and from balloons and dirigibles which observe the enemy's movements and detect his positions.

> Alfred von Schlieffen, *Cannae*, 1913

Generalship, at least in my case, came not by instinct, unsought, but by understanding, hard study and brain concentration. Had it come easy to me, I should not have done it as well.

> T.E. Lawrence, 1888–1935, *Seven Pillars of Wisdom*, 1935

Efficiency in a general, his soldiers have a right to expect; geniality they are usually right to suspect.

> Sir Archibald Wavell, *Generals and Generalship*, 1939

An important difference between a military operation and a surgical operation is that the patient is not tied down. But it is a common fault of generalship to assume that he is.

B.H. Liddell Hart, *Thoughts on War*, 1944

Two fundamental lessons of war experience are – never to check momentum; never to resume mere pushing.

B.H. Liddell Hart, *Thoughts on War*, 1944

A vital faculty of generalship is the power of grasping instantly the picture of the ground and the situation, of relating the one to the other, and the local to the general.

B.H. Liddell Hart, *Thoughts on War*, 1944

The acid test of an officer who aspires to high command is his ability to be able to grasp quickly the essentials of a military problem, to decide rapidly what he will do, to make it quite clear to all concerned what he intends to achieve and how he will do it, and then to see that his subordinate commanders get on with the job.

Montgomery of Alamein, *Memoirs*, 1958

The capacity to understand the workings of the other man's mind is an essential element in generalship.

John Connell, *Wavell, Soldier and Scholar*, 1964

Glory

There must be a beginning of any great matter, but the continuing unto the end until it be thoroughly finished yields the true glory.

Sir Francis Drake, to Lord Walsingham, 17 May 1587

To overcome in battle and subdue
Nations, and bring home spoils with infinite
Manslaughter, shall be the highest pitch
Of human glory.

John Milton, *Paradise Lost*, 1667

I am envious only of glory; for if it be a sin to covet glory, I am the most offending soul alive.

> Horatio Nelson, letter to Lady Hamilton, 18 February 1800

The love of glory can only create a great hero, the contempt of it creates a great man.

> Charles Maurice Talleyrand, 1754–1838

> We carved not a line, and we raised not a stone
> But we left him alone with his glory.
>> Charles Wolfe, *The Burial of Sir John Moore*, 1817

Military honour and glory have declined steadily since the destruction of the feudal system, which assured pre-eminence to men-at-arms.

> Napoleon I, *Political Aphorisms*, 1848

The love of glory, the ardent desire for honourable distinction by honourable deeds, is among the most potent and elevating of military motives.

> Alfred Thayer Mahan, *Life of Nelson*, 1897

Guerrilla Warfare

Militia and armed civilians cannot and should not be employed against the main force of the enemy, or even against sizeable units. They should not try to crack the core, but only nibble along the surface and on the edges. They should rise in provinces lying to one side of the main theatre of war, which the invader does not enter in force, in order to draw these areas entirely from his grasp. These storm clouds, forming on his flanks, should also follow to the rear of his advance.

> Karl von Clausewitz, *On War*, 1832

Air power may be effective against elaborate armies: but against irregulars it has no more than moral value.

> T.E. Lawrence, letter to A.P. Wavell, 21 May 1923

Guerrilla war is far more intellectual than a bayonet charge.

> T.E. Lawrence, 1888–1935, *The Science of Guerrilla Warfare*

The few active rebels must have the qualities of speed and endurance, ubiquity and independence of arteries of supply. They must have the technical equipment to destroy or paralyse the enemy's organized communications.

> T.E. Lawrence, 1888–1935, *The Science of Guerrilla Warfare*

Guerrilla war must have a friendly population, not actively friendly, but sympathetic to the point of not betraying rebel movements to the enemy. Rebellions can be made by two per cent active in a striking force, and 98 per cent passively sympathetic.

> T.E. Lawrence, 1888–1935, *The Science of Guerrilla Warfare*

... innumerable gnats, which, by biting a giant both in front and in rear, ultimately exhaust him. They make themselves as unendurable as a group of cruel and hateful devils, and as they grow and attain gigantic proportions, they will find that their victim is not only exhausted but practically perishing.

> Mao Tse-tung, *On Guerrilla Warfare*, 1937

Many people think it is impossible for guerrillas to exist for long in the enemy's rear. Such a belief reveals lack of comprehension of the relationship that should exist between the people and the troops. The former may be likened to water and the latter to the fish who inhabit it.

> Mao Tse-tung, *On Guerrilla Warfare*, 1937

The ability to run away is the very characteristic of the guerrilla.

> Mao Tse-tung, *Strategic Problems in the Anti-Japanese Guerrilla War*, 1939

When the enemy advances, we retreat.
When he escapes we harass.
When he retreats we pursue.
When he is tired we attack.
When he burns we put out the fire.
When he loots we attack.
When he pursues we hide.
When he retreats we return.

> Mao Tse-tung, 1893–1976 (after Sun Tzu 400–320 BC)

Select the tactic of seeming to come from east and attacking from the west; avoid the solid, attack the hollow; attack, withdraw; deliver a lightning blow, seek a lightning decision ... Withdraw when he advances; harass him when he stops; strike him when he is weary; pursue him when he withdraws.

> Mao Tse-tung, *On Guerrilla War*, 1937

Guerrilla war is a kind of war waged by the few but dependent on the support of the many.

> B.H. Liddell Hart, foreword to *Guerrilla Warfare*, 1961

Liberation wars will continue to exist as long as imperialism exists ... Such wars are not only admissible, but inevitable ... We recognize such wars, and we will help the peoples striving for their independence. The Communists fully support such wars and march in the front rank with peoples waging liberation struggles.

> Nikita Khruschev, speech, 1961

There is another type of warfare – new in its intensity, ancient in its origin – war by guerrillas, subversives, insurgents, assassins; war by ambush instead of by combat, by infiltration instead of aggression, seeking victory by eroding and exhausting the enemy instead of engaging him. It is a form of warfare uniquely adapted to what have been strangely called 'wars of liberation', to undermine the efforts of new and poor countries to maintain the freedom that they have finally achieved. It preys on unrest and ethnic conflicts.

> President John F. Kennedy, address to the US Naval Academy, 6 June 1962

Guns

Don't forget your great guns, which are the most respectable arguments of the rights of kings.

> Frederick the Great, letter to his brother, Prince Henry, 21 April 1759

As a general rule, the maxim of marching to the sound of the guns is a wise one.

> Antoine Henri Jomini, *Political and Military Summary of the Campaign of 1815*, 1839

Guns will make us powerful; butter will only make us fat.
>Herman Goering, radio broadcast, summer of 1936, often misquoted as 'Guns before butter'.

Every Communist must grasp the truth, 'Political power grows out of the barrel of a gun'.
>Mao Tse-tung, *Selected Works*, 1961, vol.ii, *Problems of War and Strategy*, 1938

Gustavus Adolphus II (1594–1632)

Consider the great Gustavus Adolphus! In eighteen months he won one battle, lost a second, and was killed in the third. His fame was won at a bargain price.
>Napoleon I, letter to General Gaspard Gourgaud, St Helena, 1818

Hannibal (247–183 BC)

... the most audacious of all, probably the most stunning, so
hardy, so sure, so great in all things.
Napoleon I, 1769–1821

... by many degrees the greatest soldier on record.
The Duke of Wellington, 1769–1852

Harbour

I am quite homeless. I cannot reach Germany; we possess no
other secure harbour. I must plough the seas of the world
doing as much mischief as I can till my ammunition is
exhausted or until a foe far superior in power succeeds in
catching me.
Vice-Admiral Graf von Spee, 1914, quoted in papers
belonging to Vice-Admiral K.G.B. Dewar

Headquarters

... a house with roomy offices, where telegraph and wireless,
telephone and signalling instruments are at hand, while a fleet
of automobiles and motorcycles, ready for the longest trips,
wait for orders. Here, in a comfortable chair before a large table,
the modern Alexander overlooks the whole battlefield on a
map. From here he telephones inspiring words, and here he
receives the reports from army and corps commanders and
from balloons and dirigibles which observe the enemy's
movements and detect his position.
Alfred von Schlieffen, *Cannae*, 1913

Notice in [German Army] headquarters the presence of the King; the Commander-in-Chief of the Allied Forces with his general staff; of the German Princes; and also of the Minister of War, of the Minister for Foreign Affairs, and of the Federal Chancellor. There, indeed, you have an example of the command of nations going to war. The whole power of the Government accompanies the Commander-in-Chief, in order to put at his disposition all the resources of diplomacy, of finance, and of the national soil; in order that the military enterprise to which the nation has given all its energies ... may proceed.

Ferdinand Foch, *Precepts*, 1919

Hero

See the conquering hero comes!
Sound the trumpets, beat the drums!
Thomas Morell, 1703–84, *Joshua*

Heroism is the brilliant triumph of the soul over the flesh – that is to say, over fear ... Heroism is the dazzling and brilliant concentration of courage.
Henri-Frèderic Amiel, *Journal*, 1 October 1849

Wars may cease, but the need for heroism shall not depart from the earth, while man remains man and evil exists to be redressed.
Alfred Thayer Mahan, *Life of Nelson*, 1897

History, Military

It is as impossible to write well on the operations of war, if a man has no experience on active service, as it is to write well on politics without having been engaged in political transactions and vicissitudes.
Polybius, 200–118 BC, *Histories*

If we only act for ourselves, to neglect the study of history is not prudent; if we are entrusted with the care of others it is not just.

Samuel Johnson, 1709–84

Don't you think, Madam, that it is pleasanter to read history than to live it? Battles are fought and towns taken in every page, but a campaign takes six or seven months to hear, and achieves no greater matter at last. I dare say Alexander seemed to the coffee-houses of Pella a monstrous while about conquering the world.

Horace Walpole, letter to the Countess of Ossory, 8 October 1777

Only the study of military history is capable of giving those who have no experience of their own a clear picture of what I have just called the friction of the whole machine.

Karl von Clausewitz, *Principles of War*, 1812

The history of a battle is not unlike the history of a ball. Some individuals may recollect all the little events of which the great result is the battle won or lost; but no individual can recollect the order in which, or the exact moment at which, they occurred, which makes all the difference ... But if a true history is written, what will become of the reputation of half of those who have acquired reputations, and who deserve it for their gallantry, but who, if their mistakes and casual misconduct were made public, would not be so well thought of?

The Duke of Wellington, letter to an acquaintance, 8 August 1815

What experience and history teach is this – that peoples and governments never have learned anything from history or acted on principles derived from it.

Georg Wilhelm Hegel, *The Philosophy of History*, 1827

Generals-in-Chief must be guided by their own experience, or their genius. Tactics, evolutions, the duties and knowledge of an engineer of artillery officer, may be learned in treatises, but the science of strategy is only to be acquired by experience and by studying the campaigns of all the great captains. Gustavus Adolphus, Turenne, and Frederick, as well as Alexander, Hannibal, and Caesar, have all acted upon the same principles. These have been – to keep their forces united; to leave no weak part unguarded; to seize with rapidity on important points. Such are the principles which lead to victory, and which, by inspiring terror at the reputation of your arms, will at once maintain fidelity and secure reputation.

Napoleon I, *Maxims of War*, 1831

Read and re-read the campaigns of Alexander, Hannibal, Caesar, Gustavus Adolphus, Turenne, Eugene, and Frederick. Make them your models. This is the only way to become a great captain and to master the secrets of the art of war.

Napoleon I, *Maxims of War*, 1831

Correct theories, founded upon right principles, sustained by actual events of wars, and added to accurate military history, will form a true school of instruction for generals.

Antoine Henri Jomini, *Summary of the Art of War*, 1838

Military History, accompanied by sound criticism, is indeed the true school of war.

Antoine Henri Jomini, *Summary of the Art of War*, 1838

The study of history lies at the foundation of all sound military conclusions and practice.

Alfred Thayer Mahan, 1840–1914

The value of history in the art of war is not only to elucidate the resemblance of past and present, but also their essential differences.

Sir Julian Corbett, 1854–1922

A great war does not kill the past, it gives it new life. It may seem a catastrophe which renders all that went before insignificant and not worthy study for men of action ... But it is not so. As time gives us distance we see the flood as only one more pool in the river as it flows down to eternity.

Sir Julian Corbett, 1854–1922, *The Revival of Naval History*

Those who cannot remember the past are condemned to repeat it.

George Santayana, *The Life of Reason*, 1906

127

It is only possible to probe into the mind of a commander through historical examples.

> B.H. Liddell Hart, *Strategy*, 1929

The real way to get value out of the study of military history is to take particular situations, and as far as possible get inside the skin of the man who made a decision, realize the conditions in which the decision was made, and then see in what way you could have improved on it.

> Sir Archibald Wavell, lecture to Officers, Aldershot,
> c.1930

Study the human side of military history, which is not a matter of cold-blooded formulas or diagrams, or nursery-book principles, such as: Be good and you will be happy. Be mobile and you will be victorious. Interior lines at night are the general's delight. Exterior lines at morning are the general's warning. And so on.

> Sir Archibald Wavell, lecture to Officers, Aldershot,
> c. 1930

In battle, two things are usually required of the Commander-in-Chief: to make a good plan for his army and, secondly, to keep a strong reserve ... But in order to make his plan, the General must not only reconnoitre the battle-ground, he must also study the achievements of the great Captains of the past.

> Sir Winston Churchill, *Painting as a Pastime*, 1932

It is true that full study of war will not seriously assist a subaltern on picket duty; but when it comes to understanding present war conditions and the probable origins of the next war, a deep and impartial knowledge of history is essential. Further still, as it is not subalterns or generals who make wars, but governments and nations, unless the people as a whole have some understanding of what war meant in past ages, their opinions on war ... today will be purely alchemical.

> J.F.C. Fuller, *Decisive Battles*, 1939

Though the military art is essentially a practical one, the opportunities of practising it are rare. Even the largest-scale peace manoeuvres are only a feeble shadow of the real thing. So that a soldier desirous of acquiring skill in handling troops is forced to theoretical study of Great Captains.

> Sir Archibald Wavell, lecture to Officers, Aldershot, c.1930

For my strategy, I could find no teachers in the field: but behind me there were some years of military reading, and even in the little that I have written about it, you may be able to trace the allusions and quotations, the conscious analogies.

> T.E. Lawrence, 1888–1935

War is the highest form of struggle between nations, and thus the study of military matters brooks not a moment's delay, and must be learned not only by our commanders, but also by members of the Party.

> Mao Tse-tung, *On the Study of War*, 1936

It should be the duty of every soldier to reflect on the experiences of the past, in the endeavour to discover improvements, in his particular sphere of action, which are practicable in the immediate future.

> B.H. Liddell Hart, *Thoughts on War*, 1944

The practical value of history is to throw the film of the past through the material projector of the present onto the screen of the future.

> B.H. Liddell Hart, *Thoughts on War*, 1944

Most official accounts of past wars are deceptively well written, and seem to omit many important matters – in particular, anything which might indicate that any of our commanders ever made the slightest mistake. They are therefore useless as a source of instruction.

> Montgomery of Alamein, *Memoirs*, 1958

Hitler, Adolf (1899–45)

Adolf Hitler is a bloodthirsty guttersnipe, a monster of wickedness, insatiable in his lust for blood and plunder.
> Sir Winston Churchill, on the German invasion of Russia, 22 June 1941

Honour

Better a thousand times to die with glory than live without honour.
> Louis VI of France, 1081–1137

My honour is dearer to me than my life.
> Cervantes, *Don Quixote*, 1604

Mine honour is my life; both grow in one;
Take honour from me and my life is done.
> William Shakespeare, *King Richard II*, 1595

Can honour set a leg? no: or take away the grief of a wound? no. Honour hath no skill in surgery, then? no. What is honour? a word. What is that word honour? air. Who hath it? he that died o'Wednesday. Doth he feel it? no. Doth he hear it? no. It is insensible, then? Yes, to the dead. But will it not live with the living? no. Therefore, I'll none of it. Honour is a mere scrutcheon, and so ends my catechism.
> William Shakespeare, *I King Henry IV*, 1597

If it be a sin to covet honour,
I am the most offending soul alive.
> William Shakespeare, *King Henry V*, 1598

Set honour in one eye and death i'the other
And I will look on both indifferently;
For let the gods so speed me as I love
The name of honour more than I fear death.
> William Shakespeare, *Julius Caesar*, 1599

Life every man holds dear; but the brave man
Holds honour far more precious-dear than life.
William Shakespeare, *Troilus and Cressida*, 1601

If I lose mine honour, I lose myself.
William Shakespeare, *Antony and Cleopatra*, 1606

I could not love thee, dear, so much,
Loved I not honour more.
Richard Lovelace, 1618–58, *To Lucasta, on Going to the Wars*

The first object which a general who gives battle should consider is the glory and honour of his arms; the safety and conservation of his men is but secondary.
Napoleon I, *Maxims of War*, 1831

Duty, honour, country.
Motto of the US Military Academy, West Point

Indiscipline

We started with an army in the highest order, and up to the day of battle nothing could get on better; but that event has, as usual, totally annihilated all order and discipline. The soldiers of the army have got among them about a million sterling in money ... The night of the battle, instead of being passed in getting rest and food to prepare them for the pursuit of the following day, was passed by the soldiers in looking for plunder. The consequence was, that they were incapable of marching in pursuit of the enemy, and were totally knocked up ... I am quite convinced that we now have out of the ranks double the amount of our loss in battle; and that we have lost more men in the pursuit than the enemy have; and have never in any one day made more than an ordinary march. This is the consequence of the state of discipline of the British army. We may gain the greatest victories; but we shall do no good until we so far alter our system, as to force all ranks to perform their duty.

The Duke of Wellington, after Vitoria, 21 June 1813

Infantry

Infantry is the nerve of an army.
Francis Bacon, *Essays*, 1625

The stubborn spear-men still made good
Their dark impenetrable wood,
Each stepping where his comrade stood,
The instant that he fell.
Sir Walter Scott, *Lay of the Last Minstrel*, 1805

It all depends upon that article whether we do the business or not. Give me enough of it, and I am sure.

> The Duke of Wellington, on being asked by Thomas Creevey in Brussels, whether he would beat Napoleon, May 1815. (At the moment of reply, a redcoated infantry soldier walked by)

First there is the All-Highest [the Kaiser], then the Cavalry Officer, and then the Cavalry Officer's horse. After that there is nothing, and after nothing the Infantry Officer.

> Pre-1914 Imperial German Army apothegm

One well-known Brigadier always phrases his requirements of the ideal infantryman as 'athlete, stalker, marksman'. I always feel a little inclined to put it on a lower plane and to say that the qualities of a successful poacher, cat burglar, and gunman would content me.

> Sir Archibald Wavell, *The Training of the Army for War*, 1933

When the smoke cleared away, it was the man with the sword, or the crossbow, or the rifle, who settled the final issue on the field.

> General George C. Marshall, 1939 Staff College papers

They marched back from the battle in the way of the infantry, their feet scarcely leaving the ground, their bodies rocking mechanically from side to side as if that was the only way they could lift their legs. You could see that it required the last ounce of their mental and physical energy to move their legs at all. Yet they looked as if they could keep on moving like that for ever.

> Fred Majdalany, *The Monastery*, 1950

Initiative

A general who stands motionless to receive his enemy, keeping strictly on the defensive, may fight ever so bravely, but he must give way when properly attacked.

> Antoine Henri Jomini, *Summary of the Art of War*, 1838

133

No military leader is endowed by heaven with an ability to seize the initiative. It is the intelligent leader who does so after a careful study and estimate of the situation and arrangement of the military and political factors involved.

Mao Tse-tung, *On Guerrilla War*, 1937

When so much was uncertain, the need to recover the initiative glared forth.

Sir Winston Churchill, *Their Finest Hour*, 1949 (on the situation after Dunkirk, 1940)

Innovation

Inventions do not make their first bow to armies on the battlefield. They have been in the air for some time; hawked about the ante-chambers of the men of the hour; spat upon by common sense; cold-shouldered by interests vested in what exists; held up by stale functionaries to whom the sin against the Holy Ghost is 'to make a precedent'.

Sir Ian Hamilton, *The Soul and Body of an Army*, 1921

It is by devising new weapons, and above all by scientific leadership, that we shall best cope with the enemy's superior strength.

Sir Winston Churchill, memorandum for the War Cabinet, 3 September 1940

This helplessness of the art of war, we can now see, was due to the slowness with which armies changed away from an old conception of warfare to a new one. All the technical means for ending this helplessness were present early in the war; the gasoline engine, the caterpillar tractor, the idea of an armoured vehicle capable of crossing trenches and standing machine-gun fire, the aeroplane and the light machine gun were all available. What was not available was the idea of war as a changing art or science affected by every change in the techniques of production and transport, and inevitably out of date if these techniques were not employed to the full.

Tom Wintringham, *The Story of Weapons and Tactics*, 1943

Military history is filled with the record of military improvements that have been resisted by those who would have profited richly from them.

B.H. Liddell Hart, *Thoughts on War*, 1944

Inspection

When you are ordered to visit the barracks, I would recommend it to you to confine your inspection to the outside walls: for what can be more unreasonable to expect, that you should enter the soldier's dirty rooms, and contaminate yourself with tasting their messes? As you are not used to eat salt pork or ammunition bread, it is impossible for you to judge whether they are good or not. Act in the same manner, when you are ordered to visit the hospital.

Francis Grose, *Advice to the Officers of the British Army*, 1782

Insubordination

You know, Foley, I have only one eye; I have a right to be blind sometimes – I really do not see the signal.

Horatio Nelson, to his Flag Captain at Copenhagen, after having had his attention invited to Sir Hyde Parker's signal, 'Discontinue engagement', 2 April 1801

Insubordination may only be the evidence of a strong mind.

Napoleon I, letter from St Helena, 1817

Intelligence

If you know the enemy and know yourself, you need not fear the results of a hundred battles. If you know yourself, but not the enemy, for every victory gained you will also suffer a defeat. If you know neither the enemy nor yourself, you will succumb in every battle.

Sun Tzu, 400–320 BC, *The Art of War*

Therefore, determine the enemy's plans and you will know which strategy will be successful and which will not.

Sun Tzu, 400–320 BC, *The Art of War*, vi

Nothing is more worthy of the attention of a good general than the endeavour to penetrate the designs of the enemy.

Niccolo Machiavelli, *Discourses*, 1531

One should know one's enemies, their alliances, their resources and nature of their country, in order to plan a campaign. One should know what to expect of one's friends, what resources one has, and foresee the future effects to determine what one has to fear or hope from political manoeuvres.

Frederick the Great, *Instructions for His Generals*, 1747

Knowledge of the country is to a general what a musket is to an infantryman and what the rules of arithmetic are to a geometrician. If he does not know the country he will do nothing but make gross mistakes ... Therefore study the country where you are going to act.

Frederick the Great, *Instructions for His Generals*, 1747

Great part of the information obtained in war is contradictory, a still greater part is false, and by far the greatest part is of a doubtful character.

Karl von Clausewitz, *On War*, 1832

Nothing should be neglected in acquiring a knowledge of the geography and military statistics of other states, so as to know their material and moral capacity for attack and defence as well as the strategic advantages of the two parties. Distinguished officers should be employed in these scientific labours and should be rewarded when they demonstrate marked ability.

Antoine Henri Jomini, *Summary of the Art of War*, 1838

When I took a decision, or adopted an alternative, it was after studying every relevant – and many an irrelevant – factor. Geography, tribal structure, religion, social customs, language, appetites, standards – all were at my finger-ends. The enemy I knew almost like my own side. I risked myself among them a hundred times, to learn.

T.E. Lawrence, letter to Liddell Hart, 26 June 1933

Invasion

I know I have the body of a weak and feeble woman, but I have the heart and stomach of a king, and of a king of England too; and think foul scorn that Parma or Spain, or any prince of Europe, should dare to invade the borders of my realm.
> Queen Elizabeth I, speech to the Troops at Tilbury on the Approach of the Armada, 1588

France is invaded; I go to put myself at the head of my troops.
> Napoleon I, at Paris, 23 January 1814

Remoteness is not a certain safeguard against invasion.
> Antoine Henri Jomini, *Summary of the Art of War*, 1838

Our first line of defence against invasion must be as ever the enemy's ports.
> Sir Winston Churchill, minute to Chiefs of Staff Committee, 5 August 1940

We are waiting for the long promised invasion. So are the fishes.
> Sir Winston Churchill, broadcast to the French people, 21 October 1940

Iron Curtain

From Stettin in the Baltic to Trieste in the Adriatic, an iron curtain has descended across the continent.
> Sir Winston Churchill, speech at Westminster college, Fulton, Missouri, 5 March 1946

Jargon

A serious menace is the return of jargon, technicalities and metaphors ... they swarm everywhere, a lawless rabble of camp followers.

> Karl von Clausewitz, *On War*, 1832

Joint Planning

I am increasingly impressed with the disadvantages of the present system of having Naval, Army and Air Force officers equally represented at all points and on all combined subjects, whether in committees or in commands. This has resulted in a paralysis of the offensive spirit.

> Sir Winston Churchill, note for the Chiefs of Staff Committee, 2 March 1942

Jungle Warfare

Nothing is easier in jungle or dispersed fighting than for a man to shirk. If he has no stomach for advancing, all he has to do is to flop into the undergrowth; in retreat he can slink out of the rearguard, join up later, and swear he was the last to leave. A patrol leader can take his men a mile into the jungle, hide there, and return with any report he fancies.

> Sir William Slim, *Defeat into Victory*, 1956

Junior Ranks

Long ago I had learned that in conversation with an irate senior, a junior should confine himself to the three remarks, 'Yes, sir', 'No, sir' and 'Sorry, sir'. Repeated in the proper sequence, they will get him through the most difficult interview with the minimum discomfort.

Sir William Slim, *Unofficial History*, 1959

Kill, Killing

Nothing is ever done in this world until men are prepared to kill one another if it is not done.
> George Bernard Shaw, 1856–1950, *Major Barbara*, 1907

When you have to kill a man it costs nothing to be polite.
> Sir Winston Churchill (on the ceremonial form of the declaration of war against Japan, 8 December 1941), *The Grand Alliance*, 1950

Knowledge

No study is possible on the battlefield; one does there simply what one can in order to apply what one knows. Therefore, in order to do even a little, one has already to know a great deal and to know it well.
> Ferdinand Foch, *The Principles of War*, 1918

Leadership

Because [a good] general regards his men as infants they will march with him into the deepest valleys. He treats them as his own beloved sons and they will die with him.

> Sun Tzu, 400–320 BC, *The Art of War*, Chapter 10.

Ten good soldiers wisely led
Will beat a hundred without a head.

> Attributed to Euripides, 480–406 BC

I shall desire all and every officer to endeavour by love and affable carriage to command his soldiers, since what is done for fear is done unwillingly, and what is unwillingly attempted can never prosper.

> Earl of Essex, at Worcester, 24 September 1642

Pay well, command well, hang well.

> Sir Ralph Hopton, *Maxims for the Management of an Army*, 1643

If you choose godly, honest men to be captains of Horse, honest men will follow them.

> Oliver Cromwell, letter to Sir William Springe, September 1643

One can be exact and just, and be loved at the same time as feared. Severity must be accompanied by kindness, but this should not have the appearance of pretence, but of goodness.

> Maurice de Saxe, *Reveries*, 1732

The commander should practise kindness and severity, should appear friendly to the soldiers, speak to them on the march, visit them while they are cooking, ask them if they are well cared for, and alleviate their needs if they have any. Officers without experience in war should be treated kindly. Their good actions should be praised. Small requests should be granted and they should not be treated in an over-bearing manner, but severity is maintained about everything regarding duty.

Frederick the Great, *Instructions for His Generals*, 1747

That quality which I wish to see the officers possess, who are at the head of the troops, is a cool, discriminating judgement when in action, which will enable them to decide with promptitude how far they can go and ought to go, with propriety; and to convey their orders, and act with such vigour and decision, that the soldiers will look up to them with confidence in the moment of action, and obey them with alacrity.

The Duke of Wellington, General Order, 15 May 1811

No man can justly be called a great captain who does not know how to organize and form the character of an army, as well as to lead it when formed.

Sir William Napier, 1785–1860

In enterprise of martial kind,
When there was any fighting,
He led his regiment from behind,
He found it less exciting.

W.S. Gilbert, *The Gondoliers*, 1889

I drew these tides of men into my hands
And wrote my will across the sky in stars.

T.E. Lawrence, dedicatory verses to *Seven Pillars of Wisdom*, 1935.

It is the fact that some men possess an inbred superiority which gives them a dominating influence over their contemporaries, and marks them out unmistakably for leadership. This phenomenon is as certain as it is mysterious. It is apparent in every association of human beings in every variety of circumstances and on every plane of culture. In a school among the boys, in a college among the students, in a factory, shipyard, or mine among the workmen, as certainly as in the

Church and in the Nation, there are those who, with an assured and unquestioned title, take the leading place, and shape the general conduct.

The Lord Bishop of Durham, Walker Trust Lecture on Leadership before the University of St Andrew's, 1934

Discipline apart, the soldiers' chief cares are, first, his personal comfort, i.e., regular rations, proper clothing, good billets, and proper hospital arrangements (square meals and a square deal in fact); and secondly, his personal safety, i.e. that he shall be put into a fight with as good a chance as possible for victory and survival.

Sir Archibald Wavell, 1883–1950

The task of leadership is not to put greatness into humanity, but to elicit it, for the greatness is already there.

John Buchan, 1875–1940

The commander must try, above all, to establish personal and comradely contact with his men, but without giving away an inch of his authority.

Field Marshal Erwin Rommel, 1891–1944, *The Rommel Papers*, 1953

A piece of spaghetti or a military unit can only be led from the front end.

Lieutenant-General George Patton, in North Africa, 1942

At the top there are great simplifications. An accepted leader has only to be sure of what it is best to do, or at least have his mind made up about it. The loyalties which centre upon Number One are enormous. If he trips, he must be sustained. If he makes mistakes, they must be covered. If he sleeps, he must not be wantonly disturbed. If he is no good he must be pole-axed.

Sir Winston Churchill, *Their Finest Hour*, 1949

The first thing a young officer must do when he joins the Army is to fight a battle, and that battle is for the hearts of his men. If he wins that battle and subsequent similar ones, his men will follow him anywhere; if he loses it, he will never do any real good.

Montgomery of Alamein, *Memoirs*, 1958

Leadership is a psychological force that has nothing to do with morals or good character or even intelligence; nothing to do with ideals of idealism. It is a matter of relative will-powers, a basic connection between one animal and the rest of the herd. Leadership is a process by which a single aim and unified action are imparted to the herd. Not surprisingly it is most in evidence in times or circumstances of danger or challenge. Leadership is not imposed like authority. It is actually welcomed and wanted by the led.

> Correlli Barnett, quoted in Staff College papers

Logistics

In order to make assured conquests it is necessary always to proceed within the rules; to advance, to establish yourself solidly, to advance and establish yourself again, and always prepare to have within reach of your army your resources and your requirements.

> Frederick the Great, *Instructions for His Generals*, 1747

It is very necessary to attend to all this detail and to trace a biscuit from Lisbon into a man's mouth on the frontier and to provide for its removal from place to place by land or by water, or no military operations can be carried out.

> Attributed to the Duke of Wellington, Peninsular Campaign, 1811

What makes the general's task so difficult is the necessity of feeding so many men and animals. If he allows himself to be guided by the supply officers he will never move and his expedition will fail.

> Napoleon I, *Maxims of War*, 1831

Logistics comprises the means and arrangements which work out the plans of strategy and tactics. Strategy decides where to act; logistics brings the troops to this point.

> Antoine Henri Jomini, *Summary of the Art of War*, 1838

A general should be capable of making all the resources of the invaded country contribute to the success of his enterprise.

> Antoine Henri Jomini, *Summary of the Art of War*, 1838

We have a claim on the output of the arsenals of London as well as of Hanyang, and what is more, it is to be delivered to us by the enemy's own transport corps. This is the sober truth, not a joke.

Mao Tse-tung, *On Guerrilla Warfare*, 1937

The soldier cannot be a fighter and a pack animal at one and the same time, any more than a field piece can be a gun and a supply vehicle combined.

J.F.C. Fuller, letter to S.L.A. Marshall, *c.* 1948

In this surprising journey, nothing seemed to have been forgotten. Parker says:

We frequently marched three sometimes four days successively and halted one day. We generally began our march about three in the morning, proceeded about four leagues or four-and-a-half by day, and reached our ground about nine. As we marched through the countries of our allies, commissaries were appointed to furnish us with all manner of necessaries for man and horse; these were brought to the ground before we arrived, and the soldiers had nothing to do but to pitch their tents, boil their kettles and lie down to rest. Surely never was such a march carried on with more order and regularity and with less fatigue both to man and horse.

Sir Winston Churchill, *Marlborough, His Life and Times*, commenting on Marlborough's March to the Danube.

The more I have seen of the war the more I realize how it all depends upon administration and transportation (what our American allies call logistics). It takes little skill or imagination to see where you would like your Army and when; it takes much knowledge and hard work to know where you can place your forces and whether you can maintain them there. A real knowledge of supply and movement factors must be the basis of every leader's plan; only then can he know how and when to take risks with these factors, and battles and wars are won only by taking risks.

Sir Archibald Wavell, *The Good Soldier*, 1948

145

In modern warfare no success is possible unless military units are adequately supplied with fuel, ammunition and food and their weapons and equipment are maintained. Modern battle is characterized by resolute and dynamic actions and by abrupt changes in the situation which call for greater quantity of supplies than was the case during the Second World War. Hence the increasingly important role of logistic continuity aimed at supplying each soldier in good time with everything he needs for fulfilling his combat mission.

Colonel General Golushko, Chief of Logistic Staff
Soviet Armed Forces, 1984

Man

Man, not men, is the most important consideration.
Napoleon I, *Maxims of War*, 1831

On foot, on horseback, on the bridge of a vessel, at the moment of danger, the same man is found. Anyone who knows him well, deduces from his action in the past what his future action will be.
Ardant du Picq, 1821–70, *Battles Studies*

Man is the foremost instrument of combat.
Ardant du Picq, 1821–70, *Battle Studies*

Manoeuvre

Manoeuvres are threats; he who appears most threatening, wins.
Ardant du Picq, 1821–70, *Battle Studies*

The strength of an army, like power in mechanics, is reckoned by multiplying the mass by the rapidity; a rapid march increases the morale of an army, and increases its means of victory. Press on!
Napoleon I, *Maxims of War*, 1831

Nearly all the battles which are regarded as masterpieces of the military art, from which have been derived the foundation of states and the fame of commanders, have been battles of manoeuvre.
Sir Winston Churchill, *The World Crisis*, 1923

We must accept that the enemy will penetrate through and between our forward formations and so we must be prepared to destroy him in depth by the resolute use of mobile forces capable of concentrating sufficient firepower at the right point in time and space.

> Field Marshal Sir Nigel Bagnall, when Commander in Chief British Army of the Rhine, 1984–5

We have got to be prepared to fight a mobile battle so that we can achieve a concentration of force at the critical point.

> Field Marshal Sir Nigel Bagnall, when Commander in Chief British Army of the Rhine 1984–5

Manners

My men can hardly hold their ground. Would you object to bringing yours into line with ours? I had the pleasure of being introduced to you at Lady Palmerston's last summer.

> One British officer to another at Inkerman, 5 November 1854

But after all when you have to kill a man, it costs nothing to be polite.

> Sir Winston Churchill, *The Grand Alliance*, 1950

Marines

That twelve hundred land Souldjers be forthwith raysed, to be in readinesse, to be distributed into his Majesty's Fleet prepared for Sea Service ...

> Charles II, Order in Council, 28 October 1664, establishing the world's first permanent corps of Marines, the Royal Marines

I never knew an appeal made to them for honour, courage, or loyalty that they did not more than realize my highest expectations. If ever the hour of real danger should come to England they will be found the Country's Sheet Anchor.

> Lord St Vincent, of the Royal Marines, 1802

The Marines are properly garrisons of His Majesty's ships, and upon no pretence ought they to be moved from a fair and safe communication with the ships to which they belong.

> The Duke of Wellington, to the House of Lords, 21 April 1837

It is a Corps which never appeared on any occasion or under any circumstances without doing honour to itself and its country.

> Marquis of Anglesey, speech, 5 August 1841, at Portsmouth, England

Connected with the Navy, there is the finest body of troops in the World, and that is those gallant Marines who are ever ready to devote themselves to the interests of their country.

> Benjamin Disraeli, speech, 18 September 1879

From the Halls of Montezuma to the shores of Tripoli,
We fight our country's battles in the air, on land and sea,
First to fight for right and freedom,
And to keep our honour clean,
We are proud to claim the title of United States Marine.

> The US Marines Hymn, author unknown

Military Law

The popular conception of a court martial is half a dozen bloodthirsty old Colonel Blimps, who take it for granted that anyone brought before them is guilty ... and who at intervals chant in unison, 'Maximum penalty – death!' In reality courts martial are almost invariably consulting their manuals; so anxious to avoid a miscarriage of justice that they are, at times, ready to allow the accused any loophole of escape. Even if they do steel themselves to passing a sentence, they are quite prepared to find it quashed because they have forgotten to mark something 'A' and attach it to the proceedings.

> Sir William Slim, *Unofficial History*, 1959

Military Mind

For even soldiers sometimes think –
Nay, Colonels have been known to reason –
And reasoners, whether clad in pink,
Or red, or blue, are on the brink
(Nine cases out of ten) of treason.
Thomas Moore, 1779–1852

Nine soldiers out of ten are born fools.
George Bernard Shaw, *Arms and the Man*, 1894

The professional military mind is by necessity an inferior and unimaginative mind; no man of high intellectual quality would willingly imprison his gifts in such a calling.
H.G. Wells, *Outline of History*, 1920

The mind of the soldier, who commands and obeys without question, is apt to be fixed, drilled, and attached to definite rules.
Sir Archibald Wavell, *Generals and Generalship*, 1939

The only thing harder than getting a new idea into the military mind is to get an old one out.
B.H. Liddell Hart, *Thoughts on War*, 1944

Prejudice against innovation is a typical characteristic of an Officer Corps which has grown up in a well-tried and proven system.
Field Marshal Erwin Rommel, *The Rommel Papers*, 1953

Mistakes

It is always a bad sign in an army when scapegoats are habitually sought out and brought to sacrifice for every conceivable mistake. It usually shows something wrong in the very highest command. It completely inhibits the willingness of junior commanders to make decisions, for they will always try to get chapter and verse for everything they do, finishing up more often than not with a miserable piece of casuistry instead of the decision which would spell release.

Field Marshal Erwin Rommel, *The Rommel Papers*, 1953

Happily for the result of the battle – and for me – I was, like other generals before me, to be saved from the consequences of my mistakes by the resourcefulness of my subordinate commanders and the stubborn valour of my troops.

Sir William Slim, *Defeat into Victory*, 1956

Money

Wars are not decided exclusively by military and naval force. Finance is scarcely less important. When other things are equal it is the large purse that wins.

Alfred Thayer Mahan, *The Influence of Sea Power on History*, 1890

Montgomery of Alamein

In defeat, unbeatable; in victory, unbearable.

Edward Marsh, *Ambrosia and Small Beer*

Morale

The human heart is the starting point in all matters pertaining to war.

Maurice de Saxe: *Reveries*, 1732

Morale makes up three quarters of the game, the relative balance of man-power accounts only for the remaining quarter.

Napoleon I, 1769–1821, *Correspondence*

In war the moral is to the material as three to one.
Napoleon I, 1769–1821, *Napoleon's Military Maxims*

In war, everything depends on morale; and morale and public opinion comprise the better part of reality.
Napoleon I, 1769–1821, *Thoughts*

A cherished cause and a general who inspires confidence by previous success are powerful means of electrifying an army and are conducive to victory.
Antoine Henri Jomini, *Summary of the Art of War*, 1838

It is the morale of armies, as well as of nations, more than anything else, which makes victories and their results decisive.
Antoine Henri Jomini, *Summary of the Art of War*, 1838

Morale, only morale, individual morale as a foundation under training and discipline, will bring victory.
Sir William Slim, to the Officers, 10th Indian Infantry Division, June 1941

Very many factors go into the building-up of sound morale in an army, but one of the greatest is that the men be fully employed at useful and interesting work.
Sir Winston Churchill, *The Gathering Storm*, 1948

Loss of hope, rather than loss of life, is the factor that really decides wars, battles, and even the smallest combats. The all-time experience of warfare shows that when men reach the point where they see, or feel, that further effort and sacrifice can do no more than delay the end they commonly lose the will to spin it out, and bow to the inevitable.
B.H. Liddell Hart, *Defence of the West*, 1950

Morale is a state of mind. It is that intangible force which will move a whole group of men to give their last ounce to achieve something, without counting the cost to themselves; that makes them feel they are part of something greater than themselves. If they are to feel that, their morale must, if it is to endure – and the essence of morale is that it should endure – have certain foundations. These foundations are spiritual, intellectual, and material, and that is the order of their importance. Spiritual first, because only spiritual foundations can stand real strain. Next intellectual, because men are swayed by reason as well as feeling. Material last – important, but last – because the very highest kinds of morale are often met when material conditions are lowest.

Sir William Slim, *Defeat into Victory*, 1956

The morale of the soldier is the greatest single factor in war and the best way to achieve a high morale in wartime is by success in battle.

Montgomery of Alamein, *Memoirs*, 1958

Movement

Aptitude for war is aptitude for movement.

Napoleon I, *Maxims of War*, 1831

Movement is the safety valve of fear.

B.H. Liddell Hart: *Thoughts on War*, 1944

Napoleon Bonaparte (1769–1821)

Had I succeeded, I should have died with the reputation of the greatest man that ever lived. As it is, although I have failed, I shall be considered as an extraordinary man. I have fought fifty pitched battles, almost all of which I have won. I have framed and carried into effect a code of laws that will bear my name to the most distant posterity.

> Napoleon I, letter to Barry O'Meara, St Helena, 3 March 1817

Bonaparte I never saw; though during the Battle of Waterloo we were once, I understand, within a quarter of a mile of each other. I regret it much; for he was a most extraordinary man.

> The Duke of Wellington, quoted in Samuel Rogers, *Recollections*, 1827

I used to say of him [Napoleon] that his presence on the field made the difference of forty thousand men.

> Stanhope, *Notes of Conversations with the Duke of Wellington*, 2 November 1831

Bonaparte's whole life, civil, political, and military, was a fraud. There was not a transaction, great or small, in which lying and fraud were not introduced ... Bonaparte's foreign policy was force and menace, aided by fraud and corruption. If the fraud was discovered, force and menace succeeded.

> The Duke of Wellington, 1769–1852

He was sent into this world to teach generals and statesmen what they should avoid. His victories teach what may be accomplished by activity, boldness, and skill; his disasters what might have been avoided by prudence.

> Antoine Henri Jomini, *Summary of the Art of War*, 1838

Napoleon attempted the impossible, which is beyond even genius.

Ardant du Picq, 1821–70, Battle Studies

Nation, National

No nation need expect to be great unless it makes the study of arms its principal honour and occupation.

Francis Bacon, 1561–1626

It is a narrow policy to suppose that this country or that is to be marked out as the eternal ally or the perpetual enemy ... We have no eternal allies, and we have no perpetual enemies. Our interests are eternal and perpetual, and those interests it is our duty to follow.

Lord Palmerston, to the House of Commons, 1848

In order to assure an adequate national defence, it is necessary – and sufficient – to be in a position in case of war, to conquer the command of the air.

Giulio Douhet, The Command of the Air, 1921

No foreign policy can have validity if there is no adequate force behind it and no national readiness to make the necessary sacrifices to produce that force.

Sir Winston Churchill, The Gathering Storm, 1948

Let every nation know, whether it wishes us well or ill, that we shall pay any price, bear any burden, meet any hardship, support any friend, oppose any foe, to assure the survival and success of liberty.

President John F. Kennedy, Inaugural Address, 20 January 1961

Navy, Naval

It is upon the Navy, under the good Providence of God, that the wealth, safety and strength of the kingdom do chiefly depend.

Charles II, preamble to the Articles of War, c.1670

Hearts of oak are our ships,
Jolly tars are our men,
We are always ready, steady, boys, steady,
We'll fight and we'll conquer again and again.
>> David Garrick, *Hearts of Oak*, 1759

Ye mariners of England,
That guard our native seas;
Whose flag has braved a thousand years,
The battle and the breeze!
>> Thomas Campbell, *Ye Mariners of England*, 1800

Eternal Father! strong to save,
Whose arm hath bound the restless wave,
Who bidd'st the mighty ocean deep
Its own appointed limits keep:
Oh, hear us when we cry to Thee
For those in peril on the sea!
>> William Whiting, *Hymn for Seafarers*, 1860

Now landsmen all, whoever you may be,
If you want to rise to the top of the tree,
If your soul isn't fettered to an office stool,
Be careful to be guided by this golden rule –
Stick close to your desks and never go to sea,
And you all may be Rulers of the Queen's Navee!
>> W.S. Gilbert, HMS *Pinafore*, 1878

In war, the proper objective of the navy is the enemy's navy.
>> Alfred Thayer Mahan, *Naval Strategy*, 1911

Traditions of the Royal Navy? I'll give you traditions of the Navy – rum, buggery, and the lash.
>> Sir Winston Churchill, to the Board of Admiralty, 1939

You may look at the map and see flags stuck in at different points and consider that the results will be certain, but when you get out on the sea with its vast distances, its storms and mists, and with night coming on, and all the uncertainties which exist, you cannot possibly expect that the kind of conditions which would be appropriate to the movements of armies have any applications to the haphazard conditions of war at sea.

> Sir Winston Churchill, to the House of Commons, 11 October 1940

Men go in to the Navy ... thinking they will enjoy it. They do enjoy it for about a year, at least the stupid ones do, riding back and forth quite dully on ships. The bright ones find that they don't like it in half a year, but there's always the thought of that pension if only they stay in. So they stay ... Gradually they become crazy. Crazier and crazier. Only the Navy has no way of distinguishing between the sane and the insane. Only about 5 per cent of the Royal Navy have the sea in their veins. They are the ones who become captains. Thereafter, they are segregated on their bridges. If they are not mad before this, they go mad then. And the maddest of these become admirals.

> Attributed to George Bernard Shaw, 1856–1950

There is no blitzkrieg possible in naval warfare – no lightning flash over the seas, striking down an opponent. Seapower acts more like radium – beneficial to those who use it and are shielded, it destroys the tissues of those who are exposed to it.

> B.H. Liddell Hart, *Defence of the West*, 1950

Nelson, Horatio (1758–1805)

Before this time tomorrow I shall have gained a Peerage or Westminster Abbey.

> Horatio Nelson, before the Battle of the Nile, 1 August 1798

I have only one eye – I have a right to be blind sometimes ... I really do not see the signal!

> Horatio Nelson, at the Battle of Copenhagen, 2 April 1801

I believe my arrival was most welcome, not only to the Commander of the Fleet but almost to every individual in it; and when I came to explain to them the 'Nelson touch', it was like an electric shock. Some shed tears, all approved – 'It was new – it was singular – it was simple!' ... Some may be Judas's; but the majority are much pleased with my commanding them.

> Letter to Lady Hamilton, 1 October 1805

England expects that every man will do his duty.

> Horatio Nelson, before Trafalgar, 21 October 1805

This is too warm work, Hardy, to last long.

> Horatio Nelson, at Trafalgar, 21 October 1805

Kiss me Hardy.

> Horatio Nelson, having been mortally wounded at Trafalgar, 21 October 1805

Thank God I have done my duty.

> Horatio Nelson, Dying words in the cockpit of HMS *Victory*, 21 October 1805

Rarely has a man been more favoured in the hour of his appearing; never one so fortunate in the moment of his death.

> Alfred Thayer Mahan, *Life of Nelson*, 1897

Ney, Michel (1769–1815)

... the bravest of the brave.

> Napoleon I, 1769–1821

Night Operations

In general, I believe that night attacks are only good when you are so weak that you dare not attack the enemy in daylight.

> Frederick the Great, *Instructions for His Generals*, 1747

Darkness is a friend to the skilled infantryman.

> B.H. Liddell Hart, *Thoughts on War*, 1944

Night operations have always been considered to be an important feature in the general course of combat operations; as new areas are conquered in the sphere of technical equipment, the role of the night operations is bound to grow still greater.

M.V. Frunze, quoted in Staff College papers

Nuclear Weapons

The annihilating character of these agencies may bring an utterly unforeseeable security to mankind ... It may be ... that when the advance of destructive weapons enables everyone to kill everybody else no one will want to kill anyone at all. At any rate it seems pretty safe to say that a war which begins by both sides suffering what they dread most – and that is undoubtedly the case now – is less likely to occur than one which dangles the lurid prizes of former days before ambitious eyes.

Winston Churchill, to the House of Commons, 3 November 1953

By carrying destructiveness to a suicidal extreme, atomic power is stimulating and accelerating a reversion to the indirect methods that are the essence of strategy – since they endow war with intelligent properties that raise it above the brute application of force.

B.H. Liddell Hart, *Strategy*, 1954

The atomic bomb is a paper tiger which the US reactionaries use to scare people. It looks terrible, but in fact it isn't ... All reactionaries are paper tigers.

Mao Tse-tung, talk with Anna Louise Strong, vol. iv, August 1946

Can one guess how great will be the toll of human casualties in a future war? Possibly it would be a third of the 2,700 million inhabitants of the entire world – i.e., only 900 million people. I consider this to be even low if atomic bombs actually fall. Of course it is most terrible. But even half would not be so bad ... If a half of humanity were destroyed, the other half would remain but imperialism would be destroyed entirely and there would be only Socialism in all the world.

Mao Tse-tung, at Moscow conference, 1957

Obedience

I must, I will be obeyed.
> Sir George Rodney, letter from Gibraltar to the
> Admiralty, 28 January 1780

Discipline is summed up in one word, obedience.
> Lord St Vincent, 1735–1823

I do not believe in the proverb that in order to be able to command one must know how to obey – insubordination may only be the evidence of a strong mind.
> Napoleon I, letter from St Helena, 1817

There is nothing in war which is of greater importance than obedience.
> Karl von Clausewitz, *On War*, 1832

Soldiers must obey in all things. They may and do laugh at foolish orders, but they nevertheless obey, not because they are blindly obedient, but because they know that to disobey is to break the backbone of their profession.
> Sir Charles Napier, 1782–1853

Theirs not to make reply,
Theirs not to reason why,
Theirs but to do and die.
> Alfred Lord Tennyson, *The Charge of the Light Brigade*,
> 1854

The duty of obedience is not merely military but moral. It is not an arbitrary rule, but one essential and fundamental; the expression of a principle without which military organization would go to pieces, and military success be impossible.
> Alfred Thayer Mahan, *Retrospect and Prospect*, 1902

The efficiency of a war administration depends mainly upon whether decisions emanating from the highest approved authority are in fact strictly, faithfully, and punctually obeyed.
> Sir Winston Churchill, *Their Finest Hour*, 1949

Occupation

A capital which is occupied by an enemy is like a girl who has lost her virginity.
> Napoleon I, *Maxims of War*, 1831

Odds

CAPTAIN OF THE FLAGSHIP. There are eight sail of the line, Sir John.
LORD ST VINCENT. Very well, Sir.
CAPTAIN OF THE FLAGSHIP. There are twenty sail of the line, Sir John.
LORD ST VINCENT. Very well, Sir.
CAPTAIN OF THE FLAGSHIP. There are twenty-five sail of the line, Sir John.
LORD ST VINCENT. Very well, Sir.
CAPTAIN OF THE FLAGSHIP. There are twenty-seven sail of the line, Sir John.
LORD ST VINCENT. Enough Sir, no more of that. If there are fifty sail I will go through them.
> At the battle of Cape St Vincent, 14 February 1797;
> Lord St Vincent had fifteen ships of the line in his
> fleet.

Offensive

An offensive, daring kind of war will awe the Indians and ruin the French. Block-houses and a trembling defensive encourage the meanest scoundrels to attack us.
> James Wolfe, letter to Jeffrey Lord Amherst regarding
> the forthcoming North American campaign, 1758.

When once the offensive has been assumed, it must be sustained to the last extremity.

> Napoleon I, *Maxims of War*, 1831

The best thing for an army on the defensive is to know how to take the offensive at the proper time, and to take it.

> Antoine Henri Jomini, *Summary of the Art of War*, 1836

War, once declared, must be waged offensively, aggressively. The enemy must not be fended off, but smitten down. You may then spare him every exaction, relinquish every gain; but till down he must be struck incessantly and remorselessly.

> Alfred Thayer Mahan, *The Interest of America in Sea Power*, 1896

The role of an army is to march to the sound of the guns.

> Jean Dutourd, *Taxis of the Marne*, 1957

Officers

Among those who are placed at the head of armies, there are some who are so deeply immersed in sloth and indolence that they lose all attention both to the safety of their country and their own. Others are immoderately fond of wine, so that their senses are always disordered by it before they sleep. Others abandon themselves to the love of women – a passion so infatuating that those whom it has once possessed will often sacrifice even their honour and lives to the indulgence of it.

> Polybius, 200–118 BC, *Histories*

I beseech you to be careful what captains of Horse you choose, what men be mounted; a few honest men are better than numbers ... If you choose honest godly men to be captains of Horse, honest men will follow them ... I had rather have a plain russet-coated captain that knows what he fights for, and loves what he knows, than that which you call a gentleman and is nothing else.

> Oliver Cromwell, letter to Sir William Springe, September 1643

162

When an officer comes on parade, every man in the barrack square should tremble in his shoes.
Frederick the Great, 1712–86

You cannot take too much pains to maintain subordination in your corps. The subalterns of the British army are but too apt to think themselves gentlemen; a mistake which it is your business to rectify. Put them, as often as you can, upon the most disagreeable and ungentlemanly duties.
Francis Grose, *Advice to the Officers of the British Army*, 1782

Discipline begins in the Wardroom. I dread not the seamen. It is the indiscreet conversations of the officers and their presumptious discussion of the orders they receive that produce all our ills.
Lord St Vincent, 1735–1823

My brave officers ... Such a gallant set of fellows! Such a band of brothers! My heart swells at the thought of them!
Horatio Nelson, 1758–1805

It is singular how a man loses or gains caste with his comrades from his behaviour, and how closely he is observed in the field. The officers, too, are commented upon and closely observed. The men are very proud of those who are brave in the field, and kind and considerate to the soldiers under them. An act of kindness done by an officer has often during the battle been the cause of his life being saved ... I know from experience that in our army the men like best to be officered by gentlemen, men whose education has rendered them more kind in manners than your coarse officer, sprung from obscure origin, and whose style is brutal and overbearing. '
Recollections of Rifleman Harris, Peninsula, 1808

Officers used to serve, not for their starvation pay, but for love of their country and on the off-chance of being able to defend it with their lives ... They deliberately, indeed joyously, faced up to a life of adventure, roving, action, exile, and poverty because it satisfied and reposed their souls.
Sir Ian Hamilton, *The Soul and Body of an Army*, 1921

In my experience, based on many years' observation, officers with high athletic qualifications are not usually successful in the higher ranks.

> Sir Winston Churchill, memorandum for Secretary of State for War, 4 February 1941

Orders

Be sure to give out a number of orders. It will at least show the troops you do not forget them. The more trifling they are, the more it shows your attention to the service; and should your orders contradict one another, it will give you an opportunity of altering them, and find subject for fresh regulations.

> Francis Grose, *Advice to the Officers of the British Army*, 1782

The orders I have given are strong, and I know how my admiral will approve of them, for they are, in a great measure, contrary to those he gave me; but the Service requires strong and vigorous measures to bring the war to a conclusion.

> Horatio Nelson, letter to Collingwood, July 1795

I find few think as I do, but to obey orders is all perfection. What would my superiors direct, did they know what is passing under my nose? To serve my King and to destroy the French I consider as the great order of all, from which little ones spring; and if one of those little ones militate against it, I go back to obey the great order.

> Horatio Nelson, letter from Palermo, March 1799

I shall endeavour to comply with all their Lordships' directions in such manner as, to the best of my judgement, will answer their intentions in employing me here.

> Horatio Nelson, 1758–1805, Staff College papers.

Nobody in the British Army ever reads a regulation or an order as if it were to be a guide for his conduct, or in any other manner than as an amusing novel.

> The Duke of Wellington, 1769–1852

Order, counter-order, disorder.

Helmuth von Moltke ('The Elder') 1800–91

I give orders only when they are necessary. I expect them to be executed at once and to the letter and that no unit under my command shall make changes, still less give orders to the contrary or delay execution through unnecessary red tape.

Field Marshall Erwin Rommel, letter of instruction to subordinate commanders, 22 April 1944.

Promulgation of an order represents not over ten per cent of your responsibility. The remaining ninety per cent consists in assuring through personal supervision on the ground, by yourself and your staff, proper and vigorous execution.

General George Patton, 1885–1945

Operation orders do not win battles without the valour and endurance of the soldiers who carry them out.

Sir Archibald Wavell, *Soldiers and Soldiering*, 1953

A commander must train his subordinate commanders, and his own staff, to work and act on verbal orders. Those who cannot be trusted to act on clear and concise verbal orders, but want everything in writing, are useless.

Montgomery of Alamein, *Memoirs*, 1958

Parachute

If a man have a tent made of linen of which the apertures have all been stopped up, and it be twelve bracchia across and twelve in depth, he will be able to throw himself down from any great height without sustaining an injury.

Leonardo da Vinci, 1452–1519, notebook entry

Patriotism

He serves me most, who serves his country best.
Homer, *The Iliad*, x, *c.* 1000 BC

Sweet is the love of one's country.
Cervantes, *Don Quixote*, 1605

I do love
My country's good with a respect more tender.
More holy, and profound, than mine own life.
William Shakespeare, *Coriolanus*, 1607

Patriotism is the last refuge of a scoundrel.
Samuel Johnson, in Boswell's *Life of Johnson*, entry for 7 April 1775

Breathes there the man with soul so dead,
Who never to himself hath said,
This is my own, my native land!
Walter Scott, *Lay of the Last Minstrel*, 1805

Patriotism should be something more than just hating your neighbour as much as you love yourself.
'Saki' (H.H. Munro), 1870–1916

166

Patriotism is not enough. I must have no hatred or bitterness towards anyone.

> Edith Cavell, before her execution by the Germans,
> Brussels, 12 October 1915

Patriotism is like a plant whose roots stretch down into race and place sub-consciousness; a plant whose best nutriments are blood and tears; a plant which dies down in peace and flowers most brightly in war. Patriotism does not calculate, does not profiteer, does not stop to reason: in an atmosphere of danger the sap begins to stir; it lives; it takes possession of the soul.

> Sir Ian Hamilton, *The Soul and Body of an Army*, 1921

Patriotism is not a song in the street and a wreath on a column and a flag flying from a window ... it is a thing very holy and very terrible, like life itself. It is a burden to be borne, a thing to labour for and to suffer for and to die for; a thing which gives no happiness and no pleasantness – but a hard life and an unknown grave, and the respect and bowed heads of those that follow.

> John Masefield, 1878–67

I admire men who stand up for their country in defeat, even though I am on the other side.

> Sir Winston Churchill, *The Gathering Storm*, 1948

Ask not what your country can do for you – ask what you can do for your country.

> President John F. Kennedy, Inaugural Address, 20
> January 1961

Patrolling

All commanders therefore directed their attention to patrolling. In jungle warfare this is the basis of success. it not only gives eyes to the side that excels at it, and blinds its opponent, but through it the soldiers learns to move confidently in the elements in which he works. Every forward unit, not only infantry, chose its best men, formed patrols, trained and practised them, and then sent them out on business ... These

patrols came back to their regiments with stories of
success ... The stories lost nothing in the telling, and
there was no lack of competition for the next patrol ...
In about ninety per cent of these tiny patrol actions
we were successful. By the end of November our
forward troops had gone a long way towards getting
that individual feeling of superiority and that first
essential in the fighting man – the desire to close with
his enemy ...
Sir William Slim, *Defeat into Victory*, 1956

Pay

Without going into detail about the different rates of pay, I shall
only say that it should be ample ... Economy can only be
pushed to a certain point. It has limits beyond which it
degenerates into parsimony. If your pay and allowances for
officers will not support them decently, then you will have only
rich men who serve for pleasure or adventure, or indigent
wretches devoid of spirit.

Maurice de Saxe, *Reveries*, 1732

The soldier should not have any ready money. If he has a few
coins in his pocket, he thinks himself too much of a great lord
to follow his profession, and he deserts at the opening of the
campaign.

Frederick the Great, *Instructions for His Generals*, 1747

Always grumble and make difficulties when officers go to you
for money that is due to them; when you are obliged to pay
them endeavour to make it appear granting them a favour, and
tell them they are lucky dogs to get it.

Francis Grose, *Advice to the Officers of the British Army*,
1782

Peace

He shall judge between many peoples,
   and shall decide for strong nations afar off;
and they shall beat their swords into ploughshares,
   and their spears into pruning hooks;
nation shall not lift up sword against nation,
   neither shall they learn war any more;
but they shall sit every man under his vine and under his fig
    tree;
and none shall make them afraid.
       Isaiah 2

I will make your overseers peace
and your taskmasters righteousness.
Violence shall no more be heard in your land,
devastation or destruction within your borders;
you shall call your walls Salvation,
and your gates Praise.
       Isaiah 2

Peace is very apoplexy, lethargy; mulled, deaf, sleepy, insensible; a getter of more bastard children, than war is a destroyer of men.
       William Shakespeare, *Coriolanus*, 1607

The first and fundamental law of Nature ... to seek peace and ensue it.
       Thomas Hobbes, *Leviathan*, 1651

To be prepared for war is one of the most effectual means of preserving peace.
       George Washington, speech to Congress, 8 January 1790

God send us better times with peace in our borders and war in the enemy's country.
       Lord St Vincent, letter to Nelson, 22 September 1801

What a beautiful fix we are in now: peace has been declared.
       Napoleon I, after the Treaty of Amiens, 27 March 1802

War is on its last legs; and a universal peace is as sure as is the prevalence of civilization over barbarism, of liberal governments over feudal forms. The question for us is only how soon?

R.W. Emerson, *War*, 1849

If I must choose between peace and righteousness, I choose righteousness.

Theodore Roosevelt, 1859–1919, *Unwise Peace Treaties*

Eternal peace is a dream, and not even a beautiful one.

Helmuth von Moltke ('The Elder') letter to K.K. Bluntschli, 11 December 1880

Since wars begins in the minds of men, it is in the minds of men that the defences of peace must be constructed.

Constitution of the United Nations Educational, Scientific and Cultural Organisation (1946)

Pearl Harbour

Yesterday, December 7, 1941 – a date which will live in infamy – the United States of America was suddenly and deliberately attacked by naval and air forces of the Empire of Japan.

President Franklin D. Roosevelt, to Congress, 8 December 1941

Physical Exercise

The foundation of training depends on the legs and not the arms. All the secret of manoeuvre and combat is in the legs, and it is to the legs that we should apply ourselves.

Maurice de Saxe, *Reveries*, 1732

Is it really true that a seven-mile cross-country run is enforced upon all in this division, from generals to privates? ... It looks to me rather excessive. A colonel or a general ought not to exhaust himself in trying to compete with young boys running across country seven miles at a time. The duty of officers is no doubt to keep themselves fit, but still more to think of their men, and to take decisions affecting their safety or comfort. Who is the general of this division, and does he run the seven miles himself? If so, he may be more useful for football than for war. Could Napoleon have run seven miles across country at Austerlitz? Perhaps it was the other fellow he made run. In my experience, based on many years' observation, officers with high athletic qualifications are not usually successful in the higher ranks.

> Sir Winston Churchill, note for the Secretary of State for War, 4 February 1941

A man who takes a lot of exercise rarely exercises his mind adequately.

> B.H. Liddell Hart, *Thoughts on War*, 1944

Plans

It is a bad plan that cannot be altered.

> Publilius Syrus, first century BC, *Thoughts*

Be audacious and cunning in your plans, firm and persevering in their execution, determined to find a glorious end.

> Karl von Clausewitz, *Principles of War*, 1812

In forming the plan of a campaign, it is requisite to foresee everything the enemy may do, and be prepared with the necessary means to counteract it.

> Napleon I, *Maxims of War*, 1831

If I always appear prepared, it is because before entering on an undertaking, I have meditated for long and have foreseen what may occur. It is not genius which reveals to me suddenly and secretly what I should do in circumstances unexpected by others; it is thought and meditation.

> Napoleon I, 1769–1821

No plan survives contact with the enemy.
> Attributed to Helmuth von Moltke ('The Elder'),
> 1800–91

The stroke of genius that turns the fate of a battle? I don't believe in it. A battle is a complicated operation, that you prepare laboriously. If the enemy does this, you say to yourself I will do that. If such and such happens, these are the steps I shall take to meet it. You think out every possible development and decide on the way to deal with the situation created. One of these developments occurs; you put your plan in operation, and everyone says, 'What genius ...' whereas the credit is really due to the labour of preparation.
> Ferdinand Foch, interview, April 1919

My war was overthought, because I was not a soldier.
> T.E. Lawrence, *Seven Pillars of Wisdom*, 1926

Politicians

The politician should fall silent the moment mobilization begins, and not resume his precedence until the strategist has informed the King, after the total defeat of the enemy, that he has completed his task.
> Helmuth von Moltke ('The Elder'), 1800–91

In acquiring proficiency in his branch, the politician has many advantages over the soldier; he is always 'in the field,' while the soldier's opportunities of practising his trade in peace are few and artificial ... The politician, who has to persuade and confute, must keep an open and flexible mind, accustomed to criticism and argument; the mind of the soldier, who commands and obeys without question, is apt to be fixed, drilled, and attached to definite rules.
> Sir Archibald Wavell, *Generals and Generalship*, 1939

Politico-Military Interface

Policy is the intelligent faculty, war only the instrument, not the reverse. The subordination of the military view to the political is, therefore, the only thing possible.
Karl von Clausewitz, *On War*, 1832

When soldiers deal with one another, all goes well; but, as soon as the diplomats step in, the result is unadulterated stupidity.
Alexander II of Russia, letter, 1863

Democracy is the best system of government yet devised, but it suffers from one grave defect – it does not encourage those military virtues upon which, in an envious world, it must frequently depend for survival.
Major Guy du Maurier, 1865–1915

That the soldier is but the servant of the statesman, as war is but an instrument of diplomacy, no educated soldier will deny. Politics must always exercise an extreme influence on strategy; but it cannot be gainsaid that interference with the commanders in the field is fraught with the gravest danger.
G.F.R. Henderson, *Stonewall Jackson*, 1898

War is commonly supposed to be a matter for generals and admirals, in the camp, or at sea. It would be as reasonable to say that a duel is a matter for pistols and swords. Generals with their armies and admirals with their fleets are mere weapons by the hand of the statesman.
Sir John Fortescue, lecture, 1911

Great political results often flow from correct military action; a fact which no military commander is at liberty to ignore. He may very well not know of those results; it is enough to know that they may happen.
Alfred Thayer Mahan, *Naval Strategy*, 1911

There are some militarists who say, 'We are not interested in politics but only in the profession of arms.' It is vital that these simple-minded militarists be made to realize the relationship that exists between politics and military affairs. Military action is a method used to attain a political goal. While military affairs and political affairs are not identical, it is impossible to isolate one from the other.
Mao Tse-tung, *On Guerrilla Warfare*, 1937

I do not approve of this system of encouraging political discussion in the Army among soldiers as such ... Discussions in which no controversy is desired are a farce. There cannot be controversy without prejudice to discipline. The only sound principle is 'No politics in the Army.'

Sir Winston Churchill, note for the Secretary of State for War, 17 October 1941

The soldier often regards the man of politics as unreliable, inconstant, and greedy for the limelight. Bred on imperatives, the military temperament is astonished by the number of pretences in which the statesman has to indulge. The terrible simplicities of war contrast strongly to the devious methods demanded by the art of government. The impassioned twists and turns, the dominant concern with the effect to be produced, the appearance of weighing others in terms not of their merit but of their influence – all inevitable characteristics in the civilians whose authority rests upon the popular will – cannot but worry the professional soldier, habituated as he is to a life of hard duties, self-effacement, and respect for services rendered.

President Charles de Gaulle, 1890–1970

Power

A people and nation can hope for a strong position in the world only if national character and familiarity with war fortify each other in continual interaction.

Karl von Clausewitz, *On War*, 1832

All power corrupts, and absolute power corrupts absolutely.

Lord Acton, 1834–1902, *History of Freedom*

Political power emanates from the barrel of a gun.

Mao Tse-tung, *On Guerrilla Warfare*, 1938

174

Prayer

O Lord! Thou knowest how busy I must be this day: If I forget Thee, do not Thou forget me. March on, boys!
>Sir Jacob Astley, before the Battle of Edgehill, 1642

Come on my boys, my brave boys! Let us pray heartily and fight heavily. I will run the same hazards with you. Remember the cause is for God and yourselves, your wives and children.
>Sergeant-Major-General Philip Skipton, *Turnham Green*, 1642

Save and deliver us, we humbly beseech thee, from the hands of our enemies; that we, being armed with thy defence, may be preserved evermore from all perils, to glorify thee, who art the only giver of all victory; through the merits of thy Son, Jesus Christ our Lord. Amen.
>*The Book of Common Prayer*, 1662

Dear Lord, on the morrow pray do not let me kill anyone; and, dear Lord, pray do not let anyone kill me.
>Prayer by a member of the Clan Fraser on the eve of Culloden, 16 April 1746

There is a time to pray and a time to fight. This is the time to fight.
>John Peter Gabriel Mühlenberg, sermon at Woodstock, Virginia, 1775

May the great God, whom I worship, grant to my country and for the benefit of Europe in general, a great and glorious victory, and may no misconduct in anyone tarnish it, and may humanity after the victory be the predominant feature of the British fleet.
>Horatio Nelson, prayer recorded in his diary before Trafalgar, 21 October 1805

Prepare, Prepared

When a strong man armed keepeth his palace, his goods are in peace.
>Luke 11

If we desire to avoid insult, we must be able to repel it. If we desire to secure peace, it must be known that we are at all times ready for war.

George Washington, 1732–99

The country must have a large and efficient army, one capable of meeting the enemy abroad, or they must expect to meet him at home.

The Duke of Wellington, letter, 28 January 1811

All forms of war cannot be indiscriminately condemned; so long as there are nations and empires, each prepared callously to exterminate its rival, all alike must be equipped for war.

Sigmund Freud, 1856–1939, letter to Albert Einstein

Only when our arms are sufficient beyond doubt can we be certain that they will never be employed.

President John F. Kennedy, Inaugural Address, 20 January 1961

Press

Four hostile newspapers are more to be feared than a thousand bayonets.

Napoleon I, 1769–1821, *Maxims of War*, 1831

Anti-press, anti-publicity, Official Secrets Acts have kept step with loud professions of belief in open diplomacy and the wisdom of the sovereign people.

Sir Ian Hamilton, *The Soul and Body of an Army*, 1921

Principles of War

War should be made methodically, for it should have a definite object; and it should be conducted according to the principles and rules of the art.

Napoleon I, *Maxims of War*, 1831

The principles of war are the same as those of a siege. Fire must be concentrated at one point, and as soon as the breach is made the equilibrium is broken and the rest is nothing.
> Napoleon I, *Maxims of War*, 1831

Keeping your forces united, being vulnerable at no point, moving rapidly on important points – these are the principles which assure victory, and, with fear, resulting from the reputation of your arms, maintain the faithfulness of allies and the obedience of conquered peoples.
> Napoleon I, *Maxims of War*, 1831

There exists a small number of fundamental principles of war, which may not be deviated from without danger, and the application of which, on the contrary, has been in all times crowned with glory.
> Antoine Henri Jomini, *Summary of the Art of War*, 1838

War acknowledges principles, and even rules, but these are not so much fetters, or bars, which compel its movement aright, as guides which warn us when it is going wrong.
> Alfred Thayer Mahan, 1840–1914, *The Influence of
> Seapower Upon History*, 1890

I would give you a word of warning on the so-called principles of war, as laid down in Field Service Regulations. For heaven's sake, don't treat those as holy writ, like the Ten Commandments, to be learned by heart, and as having by their repetition some magic, like the incantations of savage priests. They are merely a set of common sense maxims, like 'cut your coat according to your cloth', 'a rolling stone gathers no moss', 'honesty is the best policy', and so forth … Clausewitz has a different set, so has Foch, so have other military writers. They are all simply common sense, and are instinctive to the properly trained soldier.
> Sir Archibald Wavell, lecture to the Officers,
> Aldershot Command, *c.* 1930

A study of the laws of war is necessary as we require to apply them to war. To learn this is no easy matter and to apply them in practice is even harder; some officers are excellent at paper exercises and theoretical discussions in the war colleges, but when it comes to battle there are those that win and those that lose.
> Mao Tse-tung, *On the Study of War*, 1936

In battle, the art of command lies in understanding that no two situations are ever the same; each must be tackled as a wholly new problem to which there will be a wholly new answer.

Montgomery of Alamein, *Memoirs*, 1958

Prisoner of War

The George Cross. Lieutenant Terence Edward Waters (463718) (deceased), The West Yorkshire Regiment (The Prince of Wales's own), attached the Gloucestershire Regiment.

Lieutenant Waters was captured subsequent to the Battle of the Imjin River, 22nd–25th April, 1951. By this time he had sustained a serious wound in the top of the head and yet another most painful wound in the arm as a result of this action.

On the journey to Pyongyang with other captives, he set a magnificent example of courage and fortitude in remaining with wounded other ranks on the march, whom he felt it his duty to care for to the best of his ability.

Subsequently, after a journey of immense hardship and privation, the party arrived at an area west of Pyongyang adjacent to PW Camp 12 and known generally as 'The Caves' in which they were held captive. They found themselves imprisoned in a tunnel driven into the side of a hill through which a stream of water flowed continuously, flooding a great deal of the floor, in which were packed a great number of South Korean and European prisoners-of-war in rags filthy, crawling with lice. In this cavern a number died daily from wounds, sickness, or merely malnutrition: They fed on two small meals of boiled maize daily. Of medical attention there was none.

Lieutenant Waters appreciated that few, if any, of his numbers would survive these conditions, in view of their weakness and the absolute lack of attention for their wounds. After a visit from a North Korean Political Officer, who attempted to persuade them to volunteer to join a prisoner-of-war group known as 'Peace Fighters' (that is, active participants in the propaganda movement against their own side) with a promise of better food, of medical treatment and other amenities as a reward for such activity – an offer that was refused unanimously – he decided to order his men to pretend to accede to the offer in an effort to save their lives. This he did, giving the necessary instructions to the senior other rank with the British party, Sergeant Hoper, that the men would go upon his order without fail.

Whilst realising that this act would save the lives of his party, he refused to go himself, aware that the task of maintaining British prestige was vested in him.

Realising that they had failed to subvert an officer with the British party, the North Koreans now made a series of concerted efforts to persuade Lieutenant Waters to save himself by joining the camp. This he steadfastly refused to do. He died a short time after.

He was a young, inexperienced officer, comparatively recently commissioned from The Royal Military Academy, Sandhurst, yet he set an example of the highest gallantry.

London Gazette, 9 April 1954

Profession of Arms

War being an occupation by which a man cannot support himself with honour at all times, it ought not to be followed as a business by any but princes or governors of commonwealths; and if they are wise men they will not suffer any of their subjects or citizens to make that their only profession.

Niccolo Machiavelli, 1496–1527

Although custom and example render the profession of arms the noblest of all, I, for my own part, who only regard it as a philosopher, value it at its proper worth, and, indeed, find it very difficult to give it a place among the honourable professions, seeing that idleness and licentiousness are the two principle motives which now attract most men to it.

René Descartes, 1596–1650, quoted in Staff College papers

A soldier's time is passed in distress and danger or in idleness and corruption.

Samuel Johnson, to James Boswell, 10 April 1778

Every man thinks meanly of himself for not having been a soldier.

Samuel Johnson, to James Boswell, 10 April 1778

Soldiering, my dear madam, is the coward's art of attacking mercilessly when you are strong, and keeping out of harm's way when you are weak.

George Bernard Shaw, *Arms and the Man*, 1894

In no event will there be money in it; but there may always be honour and quietness of mind and worthy occupation – which are far better guarantees of happiness.

Alfred Thayer Mahan, *The Navy as a Career*, 1895

The professional military mind is by necessity an inferior and unimaginative mind; no man of high intellectual quality would willingly imprison his gifts in such a calling.

H.G. Wells, *Outline of History*, 1920

I find I have like of all the soldiers of different races who have fought with me and most of those who have fought against me. This is not strange, for there is a freemasonry among fighting soldiers that helps them to understand one another even if they are enemies.

Sir William Slim, *Unofficial History*, 1959

The military profession is more than an occupation; it is a style of life.

Morris Janowitz, *The Professional Soldier*, 1960

Promotion

I have seen some extremely good colonels become very bad generals.

Maurice de Saxe, *Reveries*, 1732

Above all, be careful never to promote an intelligent officer; a brave, chuckle-headed fellow will do full as well to execute your orders. An officer, that has an iota of knowledge above the common run, you must consider as your personal enemy.

Francis Grose, *Advice to the Officers of the British Army*, 1782

If ever you wish to rise a step above your present degree, you must learn that maxim of the art of war, of currying favour with your superiors; and you must not only cringe to the commander-in-chief himself, but you must take especial care to keep in with his favourites, and dance attendance on his secretary.

> Francis Grose, *Advice to the Officers of the British Army*, 1782

Every French soldier carries a marshal's baton in his knapsack.

> Attributed to Louis XVIII of France, 1755–1824

I will never agree to the promotion of an officer who, for ten years, has not been under fire.

> Napoleon I, 1769–1821 (O'Meara, *Napoleon in Exile*, 1822)

We are now at war, fighting for our lives, and we cannot afford to confine Army appointments to persons who have excited no hostile comment in their career.

> Sir Winston Churchill, note to the Chief of the Imperial General Staff, 19 October 1940

The value of 'tact' can be over-emphasized in selecting officers for command: positive personality will evoke a greater response than negative pleasantness.

> B.H. Liddell Hart, *Thoughts on War*, 1944

More and more does the 'System' tend to promote to control, men who have shown themselves efficient cogs in the machine ... There are few commanders in our higher commands. And even these, since their chins usually outweigh their foreheads are themselves outweighed by the majority – of commanders who are essentially staff officers.

> B.H. Liddell Hart, *Thoughts on War*, 1944

The man who never does more than supinely pass on the opinion of his seniors is brought to the top, while the really valuable man, the man who accepts nothing ready-made but has an opinion of his own, gets put on the shelf.

> Field Marshal Erwin Rommel, *The Rommel Papers*, 1953

An extensive use of weedkiller is needed in the senior ranks after a war; this will enable the first class young officers who have emerged during the war to be moved up.

Montgomery of Alamein, *Memoirs*, 1958

Propaganda

Great captains have always published statements for the benefit of the enemy, that their own troops were very strong in numbers; while to their own people, the enemy was represented as very inferior.

Napoleon I, 1769–1821, *Thoughts*

Propaganda, as inverted patriotism, draws nourishment from the sins of the enemy. If there are no sins, invent them! The aim is to make the enemy appear so great a monster that he forfeits the rights of a human being.

Sir Ian Hamilton, *The Soul and Body of an Army*, 1921

[Propaganda] is that branch of the art of lying which consists in very nearly deceiving your friends without quickly deceiving your enemies.

H.M. Cornford, *Microcosmographica Academica*, 1922

The printing press is the greatest weapon in the armoury of the modern commander.

T.E. Lawrence, *Seven Pillars of Wisdom*, 1935

… economical with the truth.

Sir Robert Armstrong, in describing a previous statement by himself, when giving evidence in Australia on the book *Spycatcher*, 1988

Psychological Warfare

Four elements make up the climate of war: danger, exertion, uncertainty, and chance.

Karl von Clausewitz, *On War*, 1832

We had to arrange their minds in order of battle just as carefully and as formally as other officers would arrange their bodies. And not only our own men's minds, though naturally they came first. We must also arrange the minds of the enemy, so far as we could reach them; then those other minds of the nation supporting us behind the firing line, since more than half the battle passed there in the back; then the minds of the enemy nation waiting the verdict; and the neutrals looking on; circle beyond circle.

> T.E. Lawrence, *Seven Pillars of Wisdom*, 1935

The place of artillery preparation for frontal attack by the infantry in trench warfare will in future be taken by revolutionary propaganda, to break down the enemy psychologically before the armies begin to function at all.

> Adolf Hitler, 1889–1945, *Mein Kampf*

Mental confusion, contradiction of feeling, indecisiveness, panic: these are our weapons.

> Adolf Hitler, to Herman Rauschning, 1939

The Red Army fights not merely for the sake of fighting but in order to conduct propaganda among the masses, arm them, and help them to establish revolutionary political power.

> Mao Tse-tung, *Manifestations of Various Non-Proletarian Ideas in the Party Organization of the Fourth Army*, 1929

Kill one, frighten ten thousand.

> Mao Tse-tung, after Sun Tzu

Quartermasters

If [the general] allows himself to be guided by the supply officers, he will never move, and his expeditions will fail.

> Napoleon I, 1769–1821, *Maxims of War*

If quartermasters and civilian officials are left to take their own time over the organization of supplies, everything is bound to be very slow. Quartermasters often tend to work by theory and base all their calculations on precedent, being satisfied if their performance comes up to the standard which this sets. This can lead to frightful disasters when there is a man on the other side who carries out his plan with greater drive and thus greater speed. In this situation the commander must be ruthless in his demands for an all out effort.

> Field Marshal Erwin Rommel, 1891–1944, *The Rommel Papers*

Quebec (13 September 1759)

Let not my brave soldiers see me drop –
The day is ours – Keep it.

> James Wolfe, on being mortally wounded during the battle of Quebec, 13 September 1759

What, do they run already? Then I die happy.

> James Wolfe, last words, on the Plains of Abraham, Quebec, 13 September 1759

Now, God be praised, I will die in peace.

> James Wolfe, last words, on the Plains of Abraham, Quebec, 13 September 1759. Both quotations are attributed to Wolfe as his last words. Perhaps he said both but this is more likely to be an example of how quotations are corrupted and added to through history.

Rank

Render therefore to all their dues; tribute to whom tribute is
due; custom to whom custom; fear to whom fear; honour to
whom honour.
Romans 13

An' two men ride a horse, one must ride behind.
William Shakespeare, Much Ado About Nothing, 1598

The barrier of rank is the highest of all barriers in the way of
access to the truth.
B.H. Liddell Hart, Thoughts on War, 1944

Every officer has his ceiling in rank, beyond which he should
not be allowed to rise – particularly in war-time.
Montgomery of Alamein, Memoirs, 1958

Rations

No soldier can fight properly unless he is properly fed on beef
and beer.
Attributed to the Duke of Marlborough, 1650–1722

Understand that the foundation of an army is the belly. It is
necessary to procure nourishment for the soldier wherever you
assemble him and wherever you wish to conduct him. This is
the primary duty of a general.
Frederick the Great, Instructions for His Generals, 1747

As it is the business of a good non-commission officer to be active in taking up all deserters, when on the march, or at any other time, you observe any ducks, geese, or fowls that have escaped the bounds of their confinement, immediately apprehend them, and take them along with you, that they may be tried for their offence at a proper season. This will prevent the soldiers from marauding.

Francis Grose, *Advice to the Officers of the British Army*, 1782

Articles of provision are not to be trifled with, or left to chance; and there is nothing more clear than that the subsistence of the troops must be certain upon the proposed service, or the service must be relinquished.

The Duke of Wellington, despatch, 18 February 1801

An army marches on its stomach.

Attributed to Napoleon I, 1769–1821

Generals should mess with the common soldiers. The Spartan system was a good one.

Napoleon I, to General Gaspard Gourgaud, St Helena, 1818

Readiness

It is a doctrine of war not to assume the enemy will not come, but rather to rely on one's readiness to meet him; not to presume that he will not attack, but rather to make one's self invincible.

Sun Tzu, 400–320 BC, *The Art of War*, viii

An army should be ready, every day, every night, and at all times of the day and night, to give all the resistance of which it is capable ... The soldier should always be furnished completely with arms and ammunition; the infantry should never be without its artillery, its cavalry, and its generals; and the different divisions of the army should be constantly ready to support, to be supported, and to protect themselves.

Napoleon I, *Maxims of War*, 1831

There has been a constant struggle on the part of the military element to keep the end – fighting, or readiness to fight – superior to mere administrative considerations ... the military man, having to do the fighting, considers that the chief necessity; the administrator equally naturally tends to think the smooth running of the machine the most admirable quality.

> Alfred Thayer Mahan, *Naval Administration and Warfare*, 1903

It is a law of life that has yet to be broken that a nation can only earn the right to live soft by being prepared to die hard in defence of its living.

> Sir Archibald Wavell, *Other Men's Flowers*, 1945

Reconnaissance

Agitate the enemy and ascertain the pattern of his movement. Determine his dispositions and so ascertain the field of battle. Probe him and learn where his strength is abundant and where deficient.

> Sun Tzu, 400–320 BC, *The Art of War*, vi

... skilfully reconnoitring defiles and fords, providing himself with trusty guides, interrogating the village priest and the chief of relays, quickly establishing relations with the inhabitants, seeking out spies, seizing letters.

> Napoleon I, *Maxims of War*, 1831

Time spent on reconnaissance is seldom wasted.

> *British Army Field Service Regulations*, 1912

In order to conquer that unknown which follows us until the very point of going into action, there is only one means, which consists in looking out until the last moment, even on the battlefield, for information.

> Ferdinand Foch, *Precepts*, 1919

187

Recruits, Recruiting

These men, as soon as enlisted, should be taught to work on entrenchments, to march in ranks, to carry heavy burdens, and to bear the sun and dust. Their meals should be coarse and moderate; they should be accustomed to lie sometimes in the open air and sometimes in tents. After this they should be instructed in the use of their arms. And if any long expedition is planned, they should be encamped as far as possible from the temptations of the city.

> Vegetius, *The Military Institution of the Romans*, i, AD 378

In respect to recruiting the army, my own opinion is, that the government have never taken an enlarged view of the subject. It is expected that people will become soldiers in the line, and leave their families to starve ... What is the consequence? That none but the worst description of men enter the regular service.

> The Duke of Wellington, letter, 28 January 1811

People talk of [soldiers] enlisting from their fine military feeling – all stuff – no such thing. Some of our men enlist from having got bastard children – some for minor offences – many more for drink; but you can hardly conceive such a set brought together, and it really is wonderful that we should have made them the fine fellows that they are.

> The Duke of Wellington, 1769–1852

It has been said by officers enthusiastic in their profession that there are three causes which make a soldier enlist, viz, being out of work, in a state of intoxication, or, jilted by his sweetheart. Yet the incentives to enlistment, which we desire to multiply, can hardly be put by Englishmen of the nineteenth century in this form, viz, more poverty, more drink, more faithless sweethearts.

> Florence Nightingale, *Notes on Matters Affecting the Health, Efficiency, and Hospital Administration of the British Army*, 1858

When the 'arf-made recruity goes out to the East,
'E acts like a babe an' 'e drinks like a beast.

> Rudyard Kipling, *The Young British Soldier*, 1890

Body and spirit I surrendered whole
To harsh Instructors – and received a soul.
Rudyard Kipling, *The Wonder (Epitaphs of the War)*, 1919

There are those who say, 'I am a farmer,' or, 'I am a student'; 'I can discuss literature but not the military arts'. This is incorrect. There is no profound difference between the farmer and the soldier. You must have courage. You simply leave your farms and become soldiers ... When you take your arms in hand, you become soldiers; when you are organized, you become military units.
Mao Tse-tung, *On Guerrilla Warfare*, 1937

Regiment

The regiment is the family. The Colonel, as the father, should have a personal acquaintance with every officer and man, and should instil a feeling of pride and affection for himself, so that his officers and men would naturally look to him for personal advice and instruction.
W.T. Sherman, *Memoirs*, 1875

Just certain knowledge ... The men did not expect every officer to be a brilliant leader, and they strongly hoped he would not be a 'pusher', but they expected him to put more than the next man into the general reservoir of courage. They did not look to him for ringing words of inspiration, but they liked to be reminded that they were the best mob in the line. No subaltern on the Western Front had read, or heard of, Wolseley's Pocket Book, but all grew to recognise the truth which Wolseley set out; 'The soldier is a peculiar being that can alone be brought to the highest efficiency by inducing him to believe that he belongs to a regiment that is infinitely superior to the other round him.' That was the Old Army's source of strength; and that faith in the regiment could be agreed through twenty battalions with very little dilution.
E.S. Turner, *Gallant Gentlemen*

Regular Soldiers

The man who devotes himself to war should regard it as a religious order into which he enters. He should have nothing, know no concern other than his troops, and should hold himself honoured in his profession.

Maurice de Saxe, *Reveries*, 1732

Regular troops alone are equal to the exigencies of modern war, as well for defence as offense, and when a substitute is attempted it must prove illusory and ruinous.

George Washington, 1732–99

None but the worst description of men enter the regular service.

The Duke of Wellington, letter, 28 January 1811

The regular officer has the traditions of forty generations of serving soldiers behind him, and to him the old weapons are the most honoured.

T.E. Lawrence, 1888–1935, Staff College papers

I longed for more Regular troops with which to rebuild and expand the Army. Wars are not won by heroic militia.

Winston Churchill, *Their Finest Hour*, 1949

War is the professional soldier's time of opportunity.

B.H. Liddell Hart, *Defence of the West*, 1950

Reinforcements

If a little help reaches you in the action itself, it determines the turn of fortune for you. The enemy is discouraged and his excited imagination sees the help as being at least twice as strong as it really is.

Frederick the Great, *Instructions for His Generals*, 1747

A seasonable reinforcement renders the success of a battle certain, because the enemy will always imagine it stronger than it really is, and lose courage accordingly.

Napoleon I, *Maxims of War*, 1831

Reports, Confidential

Personally, I would not breed from this officer.
> Remark on a Cavalry officer's confidential report, *c.*
> 1900

Reports, Operational

Must the operational reports from the Middle East be of their present inordinate length and detail? ... I suggest that the average weekly wordage of these routine telegrams should be calculated for the last two months and Air Marshal Longmore asked to reduce them to, say, one-third their present length.
> Sir Winston Churchill, note for the Chief of Air Staff,
> 12 January 1941

Reprimand

Always use the most opprobrious epithets in reprimanding the soldiers, particularly men of good character: for these men will not in the least be hurt, as they will be conscious that they do not deserve them.
> Francis Grose, *Advice to the Officers of the British Army,*
> 1782

If I had been censured every time I have run my ship, or fleets under my command, into great danger, I should long ago have been out of the service, and never in the House of Peers.
> Horatio Nelson, letter to the Admiralty, March 1805

Reserves

The great secret of battle is to have a reserve. I always had.
> The Duke of Wellington, 1769–1852 (Stanhope, *Notes of Conversations with the Duke of Wellington*, 1831)

Fatigue the opponent, if possible, with few forces and conserve a decisive mass for the critical moment. Once this decisive mass has been thrown in, it must be used with the greatest audacity.
> Karl von Clausewitz, *Principles of War*, 1812

The reserve is a club, prepared, organized, reserved, carefully maintained with a view to carrying out the one act of battle from which a result is expected – the decisive attack.
> Ferdinand Foch, *Precepts*, 1919

There is always the possibility of accident, of some flaw in materials, present in the general's mind; and the reserve is unconsciously held to meet it.
> T.E. Lawrence, 1888–1935, *Seven Pillars of Wisdom*, 1935

I have no more reserves. The only men I have left are the sentries at my gates. I will take them with me to where the line is broken, and the last of the English will be killed fighting.
> Sir John French, on the day the Worcestershire Regiment averted total disaster for the British Expeditionary Force by their stand, 31 August 1914

It is in the use and withholding of their reserves that the great Commanders have generally excelled. After all, when once the last reserve has been thrown in, the Commander's part is played ... The event must be left to pluck and to the fighting troops.
> Sir Winston Churchill, *Painting as a Pastime*, 1932

Resolution

I have not the particular shining bauble or feather in my cap for crowds to gaze at or kneel to, but I have power and resolution for foes to tremble at.
Oliver Cromwell, 1599–1658, letter, 1653

Great extremities require extraordinary resolution. The more obstinate the resistance of an army, the greater the chances of success. How many seeming impossibilities have been accomplished by men whose only resolve was death!
Napoleon I, *Maxims of War*, 1831

With equal or even inferior power ... he will win who has the resolution to advance.
Ardant du Picq, 1821–70, *Battle Studies*

$R > MV^2$ (R = Resolution)
Ardant du Picq, 1821–70

Retirement

With some Regret I quit the active Field,
Where Glory full Reward for Life does yield.
George Farquhar, *The Recruiting Officer*, 1706

The time factor ... rules the profession of arms. There is perhaps none where the dicta of the man in office are accepted with such an uncritical deference, or where the termination of an active career brings a quicker descent into careless disregard. Little wonder that many are so affected by the sudden transition as to cling pathetically to the trimmings of the past.
B.H. Liddell Hart, *Thoughts on War*, 1944

Retreat

If an army throws away all its cannon, equipments, and baggage, and everything which can strengthen it, and can enable it to act together as a body; and abandons all those who are entitled to its protection, but add to its weight and impede its progress; it must be able to march by roads through which it cannot be followed, with any prospect of being overtaken, by an army which had not made the same sacrifice.

> The Duke of Wellington, while pursuing Soult in
> Portugal, May 1804

In a retreat, besides the honour of the army, the loss of life is often greater than in two battles.

> Napoleon I, *Maxims of War*, 1831

However skilful the manoeuvres in a retreat, it will always weaken the morale of an army, because in losing the chances of success these last are transferred to the enemy. Besides, retreats always cost more men and material than the most bloody engagements; with this difference, that in a battle the enemy's loss is nearly equal to your own – whereas in a retreat the loss is on your side only.

> Napoleon I, *Maxims of War*, 1831

Revolution

The men who lose their heads most easily, and who generally show themselves weakest on days of revolution are the military; accustomed as they are to have an organized force facing them and an obedient force in their hands, they readily become confused before the tumultuous uproar of a crowd and in the presence of hesitation and occasional connivance of their own men.

> Alexis de Tocqueville, 1805–59, *The Old Regime*

Revolution is the locomotive of history.

> Karl Marx, 1818–83, *The Communist Manifesto*, 1848

In time of revolution, with perseverance and courage, a soldier should think nothing impossible.
>Napoleon I, *Political Aphorisms*, 1848

A disorganized army and a complete breakdown of discipline has been the condition as well as the result of every victorious revolution.
>Friedrich Engels, letter to Karl Marx, 26 September 1851

Every society is pregnant with revolution, and force is the handmaiden of revolution.
>Karl Marx, 1818–83, *The Communist Manifesto*, 1848

The central task and the highest form of a revolution is to seize political power by armed force, to settle problems by war.
>Mao Tse-tung, *Problems of War and Strategy*, 1954

Riflemen

Oh Colonel Coote Manningham, He was the Man
For he invented a capital plan,
He formed a corps of riflemen
To fight for England's glory!
He dressed them all in jackets of green,
And placed them where they couldn't be seen,
And sent them in front – an invisible screen.
To fight for England's Glory!
>From the *Regimental March* of the Rifle Brigade, 'Ninety Five', adapted by William Miller from an old song, 1842

Form, Form, Riflemen Form!
Ready, be ready to meet the storm!
Riflemen, Riflemen, Riflemen form!
>Alfred Lord Tennyson, *Riflemen Form!* 1859

River Crossings

With regard to forcing a passage across rivers, I believe it is hardly possible to prevent it. A river crossing ordinarily is supported by such massive artillery fire that it is impossible to prevent an advance force from crossing, entrenching, and throwing up works to cover the bridgehead.

Maurice de Saxe, *Reveries*, 1732

The defence of a river crossing is the worst of all assignments especially if the front that you are to defend is long; in this case defence is impracticable.

Frederick the Great, *Instructions for His Generals*, 1747

Let us cross over the river, and rest under the shade of the trees.

Lieutenant General Stonewall Jackson, last words, 10 May 1863

Russia

I cannot forecast to you the action of Russia. It is a riddle wrapped in a mystery inside an enigma; but perhaps there is a key. That key is Russian national interest.

Sir Winston Churchill, radio broadcast, 1 October 1939

War to the hilt between communism and capitalism is inevitable. Today, of course, we are not strong enough to attack. Our time will come in twenty or thirty years. To win we need the element of surprise. The bourgeoisie will have to be put to sleep. So we shall begin by launching the most spectacular peace movement on record. There will be electrifying overtures and unheard of concessions, the capitalist countries, stupid and decadent, will rejoice to co-operate in their own destruction. They will leap at another chance to be friends. As soon as their guard is down, we shall smash them with our clenched fist.

Dmitri Manuilski, to the Lenin School of Political Warfare, Moscow, 1931

How can one expect any sort of respect for normal international agreements from a regime that in the thirty-seven years since the Revolution has shot as spies and traitors, amongst others, all the members of their first Inner Cabinet and all members of the party Politburo as constituted after Lenin's death except Stalin, forty-three out of fifty-three Secretaries of the Central Organization of the party, seventy out of the eighty members of the Soviet War Council, three out of every five marshals and about 60 per cent of the generals of the Soviet Army?

Sir John Slessor, *Strategy for the West*, 1954

Sacrifice

Nearby, I met Bob [the RMO] returning to the Regimental Aid Post from a talk with the Colonel. The signallers had already destroyed their sets, and Harry was stamping on the ashes of the codebook he had just burnt. We were all ready to move. In small groups, the Headquarters split up and ran over the ridge. When they had gone, I, too, came up on to the ridge crest and prepared to descend the other side. Bob was standing alone by the path that led to the steep slopes below us.

'Come on, Bob,' I said. 'We're about the last to go – you ought to have gone before this. The Colonel will be off in a minute and that will be the lot.' He looked at me for a moment before saying:

'I can't go. I must stay with the wounded.'

For a few seconds I did not comprehend his meaning: we were all making our way out – there seemed a very fair chance that some of us would make it: to stay here was to stay certainly for capture, possibly for death, when the Chinese launched their final assault on the position. And then I realised that he had weighed all this – weighed it all and made a deliberate choice: he would place his own life in the utmost jeopardy in order to remain with the wounded at the time when they would need him most. Somewhere, the words appear, 'Greater love hath no man than this ...' I knew now exactly what those words meant. Too moved to speak again, I clapped my hand upon his shoulder and went on.

A. Farrar-Hockley, *The Edge of the Sword*, 1954

Sailor

No man will be a sailor who has the contrivance to get himself in jail, for being in a ship is being in jail, with a chance of being drowned … A man in jail has more room, better food, and commonly better company.

> Samuel Johnson, to James Boswell, 16 March 1759

The sailor in a squadron fights only once in every campaign; the soldier fights every day. The sailor, whatever may be the fatigues and dangers on the sea, undergoes fewer of these than the soldier. He never suffers from hunger or thirst; he has always with him his quarters, his kitchen, his hospital and his pharmacy.

> Napoleon I, *Maxims of War*, 1831

Seapower

To be master of the sea is an abridgement of monarchy … There be many examples where sea-fights have been final to the war … But this much is certain; that he that commands the sea is at great liberty, and may take as much and as little of the war as he will. Whereas these, that be strongest by land, are many times nevertheless in great straits.

> Francis Bacon, *Of the True Greatness of Kingdoms and Estates*, 1597

Whosoever commands the sea commands that trade; whosoever commands the trade of the world commands the riches of the world, and consequently the world itself.

> Sir Walter Raleigh, *Historie of the Worlde*, 1616

The royal navy of England hath ever been its greatest defence and ornament; it is its ancient and natural strength; the floating bulwark of the island.

> William Blackstone, *Commentaries on the Laws of England*, 1765

Under all circumstances, a decisive naval superiority is to be considered a fundamental principle, and the basis upon which all hope of success must ultimately depend.

> George Washington, letter, 1780

Without a decisive naval force we can do nothing definitive, and with it everything honourable and glorious.

> George Washington, to the Marquis de la Fayette, 15 November 1781

Britannia needs no bulwarks,
No towers along the steep;
Her march is o'er the mountain waves
Her home is on the deep.

> Thomas Campbell, *Ye Mariners of England*, 1800

In the wreck of the continent, and the disappointment of our hopes there, what has been the security of this country but its naval preponderance?

> William Pitt, to the House of Commons, 2 February 1801

Had I been master of the sea, I should have been lord of the Orient.

> Napoleon I, 1769–1821 (O'Meara, *Napoleon in Exile*, 1822)

It is not the taking of individual ships or convoys, be they few or many, that strikes down the money power of a nation; it is the possession of that overbearing power on the sea which drives the enemy's flag from it, or allows it to appear only as the fugitive; and by controlling the great common, closes the highways by which commerce moves to and from the enemy's shores. This overbearing power can only be exercised by great navies.

> Alfred Thayer Mahan, *The Influence of Sea Power upon History*, 1890

Sea power in the broad sense ... includes not only the military strength afloat that rules the sea or any part of it by force of arms, but also the peaceful commerce and shipping from which alone a military fleet naturally and healthfully springs, and on which it securely rests.

> Alfred Thayer Mahan, *The Influence of Sea Power upon History*, 1890

Never did sea power play a greater or more decisive part than in the contest which determined that the course of world history would be modified by the existence of one great nation, instead of several rival states, on the North American continent.

> Alfred Thayer Mahan, *The Influence of Sea Power upon History*, 1890

The world has never seen a more impressive demonstration of the influence of sea power upon its history. Those far distant, storm-beaten ships, upon which the Grand Army never looked, stood between it and the dominion of the world.

> Alfred Thayer Mahan, *The Influence of Sea Power upon the French Revolution and Empire*, 1892 (of the Royal Navy)

The British Army should be a projectile to be fired by the British Navy.

> Sir Edward Grey, 1862–1933, in the House of Commons

When we speak of command of the seas, it does not mean command of every part of the sea at the same moment, or at every moment. It only means that we can make our will prevail ultimately in any part of the seas which may be selected for operations, and thus indirectly make our will prevail in every part of the seas.

> Sir Winston Churchill, to the House of Commons, 11 October 1940

Security

He who would preserve everything, preserves nothing.

> Frederick the Great, *Instructions for His Generals*, 1747

Ship

Being in a ship is being in jail, with the chance of being drowned.

> Samuel Johnson, to James Boswell, 16 March 1759

Chatfield, there seems to be something wrong with our bloody ships today.

> Admiral Sir David Beatty, at Jutland, to his Flag Captain after seeing HMS *Queen Mary* blow up under German gunfire, 31 May 1916

Situation, Appreciation of

With many calculations, one can win; with few one cannot. How much less chance of victory has one who make none at all! By this means I examine the situation and the outcome will be clearly apparent.

> Sun Tzu, 400–320 BC, *The Art of War*, i

The wise general in his deliberations must consider both favourable and unfavourable factors. By taking into account the favourable factors, he makes his plan feasible; by taking into account the unfavourable, he may resolve the difficulties.

> Sun Tzu, 400–320 BC, *The Art of War*, viii

A general is not easily overcome who can form a true judgement of his own and the enemy's forces.

> Vegetius, *The Military Institutions of the Romans*, AD 378

The ability of a commander to comprehend a situation and act promptly is the talent which great men have of conceiving in a moment all the advantages of the terrain and the use that they can make of it with their army.

> Frederick the Great, *Instructions for His Generals*, 1747

The one who is to draw up a plan of operations must possess a minute knowledge of the power of his adversary and of the help the latter may expect from his allies. He must compare the forces of the enemy with his own numbers and those of his allies so that he can judge which kind of war he is able to lead or to undertake.

Frederick the Great, letter, 1748

Soldiers

Soldiers fight and die to advance the wealth and luxury of the great, and they are called masters of the world without having a sod to call their own.

Tiberius Sempronius Gracchus, speech in Rome, 133 BC, Livy

He who makes war his profession cannot be otherwise then vicious. War makes thieves, and peace brings them to the gallows.

Niccolo Machiavelli, *The Art of War*, 1520

For who ought to be more faithful than a man that is entrusted with the safety of his country, and has sworn to defend it to the last drop of his blood? Who ought to be fonder of peace than those that suffer by nothing but war? Who are under greater obligations to worship God than Soldiers, who are daily exposed to innumerable dangers, and have most occasion for his protection?

Niccolo Machiavelli, *The Art of War*, preface, 1520

Soldiers in peace are like chimneys in summer.

William Cecil (Lord Burghley): Advice to His Son, *c.* 1555

… As soldiers will
That nothing do but meditate on blood,
To swearing and stern looks.

William Shakespeare, *Henry Henry V*, 1598

... A soldier,
Full of strange oaths, and bearded like the pard;
Jealous in honour, sudden and quick in quarrel,
Seeking the bubble reputation
Even in the cannon's mouth.

> William Shakespeare's, *As Your Like It*, 1599

Rude am I in my speech,
And little bless'd with the soft phrase of peace.

> William Shakespeare, *Othello*, 1604

To dare boldly,
In a fair cause, and for their country's safety:
To run upon the cannon's mouth undaunted;
To obey their leaders, and shun mutinies;
To bear with patience the winter's cold
And summer's scorching heat, and not to faint
When plenty of provision fails, with hunger,
Are the essential parts that make up a soldier.

> Philip Massinger, 1583–1640

The military pedant always talks in a camp, and is storming towns, making lodgments and fighting battles from one end of the year to the other. Everything he speaks smells of gunpowder; if you take away his artillery from him, he has not a word to say for himself.

> Joseph Addison, *The Spectator*, 30 June 1711

If my soldiers were to begin to think, not one would remain in the ranks.

> Ascribed to Frederick the Great, 1712–86

Dogs, would you live forever?

> Frederick the Great (to soldiers reluctant to advance)

Every man thinks meanly of himself for not having been a soldier, or not having been at sea.

> Samuel Johnson, to James Boswell, 31 March 1778

... a rapacious and licentious soldiery.

> Edmund Burke, to the House of Commons on Fox's
> East India Bill, 1783

WOOLWICH, February 26th, 1793

Dear Mother, Brother, Sister and aquentices,

This is the last from us in Ingland. I have just received orders for Germany under the command of the Duke of York, with 2,200 foot guards. We expect to embark tomorrow with 1 Captain, 4 Subalterns, 8 non-comishened Officers and 52 gunners to go with His Royal Highness as a Bodyguard of British Heroes. We are to lead the Dutch Prushen and Hanover Troops into the field, as there is none equal to the British Army. We are chosen troops sent by His Majesty to show an example to the other Troops, to go in front, & lead the combined army against the French which consists of 150,000 able fighten men. You may judge if we shall have anything to dow. I had the pleasure to conquer the French last war; but God knows how it will be this war. I cannot expect to escape the Bullets of my enemies much longer, as non has ever entred my flesh as yet. To be plain with you and not dishearten you, I don't ever expect to come off so cleane as I did last war. But it is death or honour. I exspeck to be a Gentleman or a Cripel. But you never shall see me to destress you. If I cannot help you, I never shall destress you.

Dear Mother, I take my family with me. Where I go, they must go. If I leave them, I should have no luck. My wife and 2 children is in good health, & I in good spirits. Fear not for us. I hope God will be on our side.

Your Loven Son & Daughter
GEO & MARY ROBERTSON

Quoted by Sir John Fortescue, *Following the Drum*, 1931

I consider nothing in this country so valuable as the life and health of the British soldier.

The Duke of Wellington in India, 1803

Fell as he was in act and mind,
He left no bolder heart behind:
Then, give him, for a soldier meet,
A soldier's cloak for winding sheet.

Walter Scott, *Rokeby*, 1813

In order to have good soldiers, a nation must always be at war.

Napoleon I, to Barry O'Meara, St Helena, 26 October 1816

The first qualification of a soldier is fortitude under fatigue and privation. Courage is only the second; hardship, poverty, and want are the best school for a soldier.

Napoleon I, *Maxims of War*, 1831

I don't know what effect these men will have on the enemy, but, by God, they frighten me.

The Duke of Wellington, on a draft of troops sent from England during the Peninsular Campaign, 1809

We have in the service the scum of the earth as common soldiers.

The Duke of Wellington, letter to Earl Bathurst after Vitoria, 2 July 1813

In India the men of the Army generally are looked upon as so many pieces of one great machine that is passive in the hands of the engineer; and as to sense or feeling, that is not thought of. The private soldier is looked upon as the lowest class of animals, and only fit to be ruled with the cat o'nine tails and the provost-sergeant. Such a course is not likely either to improve or to correct their morals, and I am sorry to say that it is very bad.

Private Robert Waterfield, 32nd Foot, 1847

Theirs not to make reply,
Theirs not to reason why,
Theirs but to do and die,

Alfred Lord Tennyson, *The Charge of the Light Brigade*, 1854

I have never been able to join in the popular cry about the recklessness, sensuality, and helplessness of the soldiers. On the contrary I should say ... that I have never seen so teachable and helpful a class as the Army generally. Give them opportunity promptly and securely to send money home and they will use it. Give them schools and lectures and they will come to use them. Give them books and games and amusements and they will leave off drinking. Give them suffering and they will bear it. Give them work and they will do it.

Florence Nightingale, letter to her sister, March 1856

The soldier as an abstract idea is a hero ... but as a social fact he is a pariah.

Article in Blackwood's Edinburgh Magazine, 1859

Tell me what find we to admire
In epaulets and scarlet coats –
In men, because they load and fire,
And know the art of cutting throats?

William Makepeace Thackeray, 1811–63, Staff College papers

[The soldier] must be taught to believe that his duties are the noblest which fall to a man's lot. He must be taught to despise all those of civil life. Soldiers, like missionaries, must be fanatics.

Sir Garnet Wolseley, The Soldier's Pocket Book, 1809

The soldier – that is, the great soldier – of today is not a romantic animal, dashing at forlorn hopes, animated by frantic sentiment, full of fancies as to a love-lady or a sovereign; but a quiet, grave man, busied in charts, exact in sums, master of the art of tactics, occupied in trivial detail; thinking, as the Duke of Wellington was once said to do, most of the shoes of his soldiers; despising all manner of éclat and eloquence; perhaps, like Count Moltke, 'silent in seven languages'.

Walter Bagehot: The English Constitution, 1867

It's Tommy this, an' Tommy that, an'
    'Chuck him out, the brute!'
But it's 'Saviour of 'is country' when the
    guns begin to shoot.

Rudyard Kipling, Tommy, 1890

An' if sometimes our conduck isn't all
    your fancy paints,
Why, single men in barricks don't grow
    into plaster saints.

Rudyard Kipling, Tommy, 1890

A soldier is an anchronism of which we must get rid.

George Bernard Shaw, The Devil's Disciple, 1897

When the military man approaches, the world locks up its spoons and packs off its womankind.
George Bernard Shaw, *Man and Superman*, 1903

You smug-faced crowds with kindling eyes
Who cheer when soldier lads march by,
Sneak home and pray you'll never know
The hell where youth and laughter go.
Siegfried Sassoon, *Pray God*, 1918

A soldier of the Great War Known unto God.
Inscription on gravestones above unidentified bodies, chosen by Kipling as literary adviser for the Imperial War Graves Commission, 1919

Boys are soldiers in their hearts already – and where is the harm? Soldiers are not pugnacious. Paul bade Timothy be a good soldier. Christ commended the centurion. Milton urged teachers to fit their pupils for all the offices of war. The very thought of danger and self sacrifice are inspirations.
Sir Ian Hamilton, *The Soul and Body of an Army*, 1921

Service fellows ... enlist mainly for a refuge against the pain of making a living.
T.E. Lawrence, letter to James Hanley, 2 July 1931

What are the qualities of the good soldier, by the development of which we make the man war-worthy – fit for any war? ... The following four – in whatever order you place them – pretty well cover the field: discipline, physical fitness, technical skill in the use of his weapons, battle-craft.
Sir Archibald Wavell, lecture at the Royal United Service Institution, 15 February 1933

It was commonly believed that any young man who joined the Army did so because he was too lazy to work, or else he had got a girl in the family way. Hardly anyone had a good word for the soldier, and mothers taught their daughters to beware of them.
Private Frank Richards, *Old Soldier Sahib*, 1936

It is not that the British soldier is braver than other soldiers. He is not – but he is brave for a bit longer, and it's that bit that counts. Endurance is the very fibre of his courage and of his character. He stays where he is until he has won. He did it at Gibraltar two hundred years ago; a few years back he was doing it at Kohima. He has done it since.

Sir William Slim, *Courage and Other Broadcasts*, 1951

Many countries produce fine soldiers, whose achievements rival those of our own. It is character that the British soldier shows beyond others that marks his greatness. Courage, endurance, skill, adaptability, discipline they may have, but none blends these qualities together as he does with this leaven gentleness and humour. Nor has any other soldier his calm unshakeable confidence of victory.

Sir William Slim, *Courage and Other Broadcasts*, 1951

The ordinary soldier has a surprisingly good nose for what is true and what false.

Field Marshal Erwin Rommel, *The Rommel Papers*, 1953

Professional soldiers are sentimental men, for all the harsh realities of their calling. In their wallets and in their memories they carry bits of philosophy, fragments of poetry, quotations from the Scriptures, which, in times of stress and danger speak to them with great meaning.

General Matthew Ridgway, US Army, *My Battles in War and Peace*, January 1956

How often have I seen them, unconscious ambassadors, showing their identity discs or photos of their wives and families, asking questions by signs, swapping cigarettes, buttons, and I am afraid, at times cap-badges.

Sir William Slim, *Unofficial History*, 1959

Soldiers, Old

There are no greater patriots than those good men who have been maimed in the service of their country.

Napoleon I, *Political Aphorisms*, 1848

Old soldiers never die:
They simply fade away.

> British soldier's song popular in World War I

He had the old soldier's knack of making himself comfortable anywhere. The rest of us used to crawl between our blankets in our shirts and trousers. Not Harry. He never slept in anything but crisp, white pyjamas. And he slept between the sheets. And in spite of the general squalor, the sheets and the pyjamas always appeared to be spotless. In the early morning he always shaved long before anyone else, and appeared spruce and fresh as if straight from a shower; his trousers were invariably well creased, and his boots shiny. How he kept this up, no one ever knew. Old soldiers have their secrets which are not divulged to lesser mortals.

> Fred Majdalany, *The Monastery*, 1950

Spanish Armada

Play out the game; there's time for that and to beat the Spanish after.

> Attributed to Sir Francis Drake, playing bowls on Plymouth Hoe, 19 July 1588, when the Spanish fleet was sighted

I know I have the body of a weak and feeble woman, but I have the heart and stomach of a king, and of a king of England, too; and think foul scorn that Parma or Spain, or any prince of Europe should dare to invade the borders of my realm.

> Elizabeth I, speech to the troops at Tilbury during the approach of the Spanish Armada, 1588

He made the wynds and waters rise
To scatter all myne enemies ...

> Elizabeth I, Songe of Thanksgiving, 1588, composed after defeat of the Armada

Staff

The futile employment ycleptd Staff should be totally done away, and all the frippery of the Army sent to the devil.
> Lord St Vincent, letter to Benjamin Tucker on naval reorganization, 1818

Great captains have no need for counsel. They study the questions which arise, and decide them, and their entourage has only to execute their decisions. But such generals are stars of the first magnitude who scarcely appear once in a century. In the great majority of cases, the leader of the army cannot do without advice. This advice may be the outcome of the deliberations of a small number of qualified men. But within this small number, one and only one opinion must prevail. The organization of the military hierarchy must ensure subordination even in thought and give the right and duty of presenting a single opinion for the examination of the general-in-chief to one man and only one.
> Helmuth von Moltke ('The Elder'), letter, 1862

The staff knew so much more of war than I did that they refused to learn from me of the strange conditions in which Arab irregulars had to act: and I could not be bothered to set up a kindergarten of the imagination for their benefit.
> T.E. Lawrence, *Seven Pillars of Wisdom*, 1926

The military staff must be adequately composed: it must contain the best brains in the fields of land, air, and sea warfare, propaganda war, technology, economics, politics and also those who know the people's life.
> Erich Ludendorff, *Total War*, 1935

My war experience led me to believe that the staff must be the servants of the troops, and that a good staff officer must serve his commander and the troops but himself be anonymous.
> Montgomery of Alamein, *Memoirs*, 1958

211

Standing Operating Procedures

Serve God daily, love one another, preserve your victuals, beware of fire, and keep good company.
> Sir John Hawkins, 1532–95, standing orders to his ships

One should, once and for all, establish standard combat procedures known to the troops, as well as to the general who leads them.
> Maurice de Saxe, *Reveries*, 1732

Thus there arises a certain methodism in warfare to take the place of art, wherever the latter is absent.
> Karl von Clausewitz, *Principles of War*, 1812

Fundamental principles of action against different arms must be laid down so definitely that complicated orders in each particular case will not be required.
> Friederich von Bernhardi, 1849–1930

Strategy

Supreme excellence consists of breaking the enemy's resistance without fighting.
> Sun Tzu, 400–320 BC, *The Art of War*, chapter 3

Those skilled in war bring the enemy to the field of battle and are not brought there by him.
> Sun Tzu, 400–320 BC, *The Art of War*, chapter 3

[Strategy] means the combination of individual engagements to attain the goal of the campaign.
> Karl von Clausewitz, *Principles of War*, 1812

The theory of warfare tries to discover how we may gain a preponderance of physical forces and material advantages at the decisive point. As this is not always possible, theory also teaches us to calculate moral factors.
> Karl von Clausewitz, *Principles of War*, 1812

Strategy is … the art of making war upon the map, and comprehends the whole of the theatre of operations.

Antoine Henri Jomini, *Summary of the Art of War*, 1838

A victory on the battlefield is of little account if it has not resulted either in breakthrough or encirclement. Though pushed back, the enemy will appear again on different ground to renew the resistance he momentarily gave up. The campaign will go on.

Alfred von Schlieffen, *Cannae*, 1913

As in a building, which however fair and beautiful the superstructure, is radically marred and imperfect if the foundation be insecure – so, if the strategy be wrong, the skill of the general on the battlefield, the valour of the soldier, the brilliancy of victory, however otherwise decisive, fail of their effect.

Alfred Thayer Mahan, *Naval Administration and Warfare*, 1903

The soundest strategy is to postpone operations until the moral disintegration of the enemy renders the delivery of the mortal blow both possible and easy.

V.I. Lenin, 1870–1924

Grand strategy must always remember that peace follows war.

B.H. Liddell Hart, *Thoughts on War*, 1944

In Napoleon, the Power Age found its prophet … Its Koran was written by Karl von Clausewitz … It has become the war creed of all nations.

J.F.C. Fuller, *Armament and History*, 1945

The true aim is not so much to seek battle as to seek a strategic situation so advantageous that if it does not of itself produce the decision, its continuation by a battle is sure to achieve this.

B.H. Liddell Hart, *Strategy*, 1954

Strength

In war, numbers alone confer no advantage. Do not advance relying on sheer military power.

Sun Tzu, 400–320 BC, *The Art of War*, ii

It is not big armies that win battles; it is the good ones.
Maurice de Saxe, *Reveries*, 1732

The first rule is to enter the field with an army as strong as possible.
Karl von Clausewitz, *On War*, 1832

The best strategy is always to be strong.
Karl von Clausewitz, *On War*, 1832

The cult of numbers is the supreme fallacy of modern warfare.
B.H. Liddell Hart, *Thoughts on War*, 1944

Submarine

Like the destroyer, the submarine has created its own type of officer and man – with language and traditions apart from the rest of the Service, and yet at heart unchangingly of the Service.
Rudyard Kipling, *The Fringes of the Fleet*, 1915

To the end that prohibition of the use of submarines as commerce destroyers shall be accepted universally as part of the law of nations, the signatory powers herewith accept that prohibition as binding between themselves, and invite all other nations to adhere thereto.
Washington Naval Treaty, 6 February 1922

It must not be forgotten that defeat of the U-boats carries with it the sovereignty of all the oceans of the world.
Sir Winston Churchill, statement to the French Admiralty, November 1939

The defeat of the U-boat is the prelude to all effective aggressive operations.
Sir Winston Churchill, to conference of the Ministers of the Crown, 11 February 1943

Surprise

The enemy must now know where I intend to give battle. For if he does not know where I intend to give battle, he must prepare in a great many places ... If he prepares to the front his rear will be weak, and if to the rear, his front will be fragile. If he prepares to the left, his right will be vulnerable and if to the right, there will be few on his left. And when he prepares everywhere he will be weak everywhere.

> Sun Tzu, 400–320 BC, *The Art of War*, vi

Everything which the enemy least expects will succeed the best.

> Frederick the Great, *Instructions for His Generals*, 1747

To be defeated is pardonable; to be surprised – never!

> Napoleon I, *Maxims of War*, 1831

War is composed of nothing but surprises. While a general should adhere to basic principles, he should never miss an opportunity to profit by such surprises. It is the essence of genius. In war there is but one favourable moment; genius grasps it.

> Napoleon I, *Maxims of War*, 1831

Surprise, the pith and marrow of war.

> Admiral of the Fleet Lord Fisher, memorandum, July 1906

Inaction leads to surprise, and surprise to defeat, which is after all only a form of surprise.

> Ferdinand Foch, *Precepts*, 1919

Movement generates surprise, and surprise gives impetus to movement.

> B.H. Liddell Hart, article 'Strategy', 1929. Based on his book *The Decisive Wars of History*, 1929

We have inflicted a complete surprise on the enemy. All our columns are inserted in the enemy's guts.

> Orde Wingate, Order of the Day to all ranks, 3rd Indian Division, 11 March 1944

Surrender

The Guard dies; it does not surrender. (La Garde meurt et ne se rend pas.)

> General Pierre de Cambronne, of the Imperial Guard at Waterloo, 18 June 1815

There is but one honourable mode of becoming prisoner of war. That is, by being taken separately; by which is meant, by being cut off, entirely, and when we can no longer make use of our weapons. In this case there can be no conditions, for honour can impose none. We yield to irresistible necessity.

> Napoleon I, *Maxims of War*, 1831

We shall never surrender.

> Sir Winston Churchill, to the House of Commons, 4 June 1940

We, the United Nations, demand from the Nazi, Fascist and Japanese tyrannies unconditional surrender. By this we mean that their will power to resist must be completely broken, and that they must yield themselves absolutely to our justice and mercy.

> Sir Winston Churchill, speech in the Guildhall, London, 30 June 1943

Nuts!

> Major General Anthony McAuliffe, US Army in reply to a German demand that he surrender the US troops in Bastogne, 23 December 1944

Sword

All they that take to the sword shall perish with the sword.

> Matthew 26

The swords of soldiers are his teeth, his fangs ...

> William Shakespeare, *King John*, 1596

They say they can obtain land and people for the King with the pen; but I say it can only be done with the sword.

> Frederick William I of Prussia, 1688–1740

The pen is mightier than the sword.

Edward George Bulwer-Lytton, 1803–73, *Richelieu*

Until the world comes to an end the ultimate decision will rest with the sword.

Kaiser Wilhelm II, speech in Berlin, 1913

Tactics

Now an army may be likened to water, for just as flowing water avoids the heights and hastens to the lowlands, so an army avoids strength and strikes weakness. And as water shapes its flow in accordance with the ground, so an army manages its victory in accordance with the situation of the enemy. And as water has no constant form, there are in war no constant conditions.

Sun Tzu, 400–320 BC, *The Art of War*, vi

Within a single square mile a hundred different orders of battle can be formed. The clever general perceives the advantages of the terrain instantly; he gains advantage from the slightest hillock, from a tiny marsh; he advances or withdraws a wing to gain superiority; he strengthens either his right or left, moves ahead or to the rear, and profits from the merest bagatelles.

Frederick the Great, *Instructions for His Generals*, 1747

What distinguishes a man from a beast of burden is thought and the faculty of bringing ideas together ... a pack mule can go on ten campaigns with Prince Eugene of Savoy and still know nothing of tactics.

Attributed to Frederick the Great, 1712–86

When you determine to risk a battle, reserve to yourself every possible choice of success, more particularly if you have to deal with an adversary of superior talent; for if you are beaten, even in the midst of your magazines and communications, woe to the vanquished!

Napoleon I, *Maxims of War*, 1831

Grand tactics is the art of posting troops upon the battlefield according to the characteristics of the ground, of bringing them into action, and of fighting them upon the ground.

Antoine Henri Jomini, *Summary of the Art of War*, 1838

The problem is to grasp, in innumerable special cases, the actual situation which is covered by the mist of uncertainty, to appraise the facts correctly and to guess the unknown elements, to reach a decision quickly and then to carry it out forcefully and relentlessly.

> Helmuth von Moltke ('The Elder'), 1800–91

Tactics is an art based on the knowledge of how to make men fight with maximum energy against fear, a maximum which organisation alone can give.

> Ardant du Picq, 1821–70, *Battle Studies*

Changes in tactics have not only taken place after changes in weapons, which necessarily is the case, but the interval between such changes has been unduly long. An improvement of weapons is due to the energy of one or two men, while changes in tactics have to overcome the inertia of a conservative class.

> Alfred Thayer Mahan, 1840–1914, *Naval Administration and Warfare*, 1903

Unless very urgent reasons to the contrary exist, strike at one end rather than at the middle, because both ends can come up to help the middle against you quicker than one end can get to help the other; and, as between the two ends, strike at the one upon which the enemy most depends for reinforcements and supplies to maintain his strength.

> Alfred Thayer Mahan, *Sea Power in its Relations with the War of 1812*, 1905

Nine-tenths of tactics are certain, and taught in books: but the irrational tenth is like the kingfisher across the pool and that is the test of generals. It can only be ensured by instinct, sharpened by thought practising the stroke so often that at the crisis it is as natural as a reflex.

> T.E. Lawrence, 1888–1935, *The Science of Guerrilla Warfare*

There is only one tactical principle which is not subject to change. It is to use the means at hand to inflict the maximum amount of wounds, death and destruction on the enemy in the minimum of time.

> General George Patton, *War As I Knew It*, 1947

The commander must decide how he will fight the battle before it begins. he must then decide how he will use the military effort at his disposal to force the battle to swing the way he wishes it to go; he must make the enemy dance to his tune from the beginning and never vice versa.

Montgomery of Alamein, *Memoirs*, 1958

Tactics is the opinion of the senior officer present.

British Army saying

Tank

A pretty mechanical toy.

Attributed to Lord Kitchener, after observing British tank test, 1915

The tank marks as great a revolution in land warfare as an armoured steamship would have marked had it appeared amongst the toilsome triremes of Actium.

Sir Ian Hamilton, *The Soul and Body of an Army*, 1921

The [tank] has been conceived, but before it could be born and waddle across no-man's land to browse upon the barbed wire of the Germans, it had first to get through the barbed wire of bureaucrats whereon fluttered still the poor rags once worn by dead inventors.

Sir Ian Hamilton, *The Soul and Body of an Army*, 1921

Where tanks are, is the front ... Wherever in future wars the battle is fought, tank troops will play the decisive role.

Heinz Guderian, *Achtung! Panzer!* 1937

With the development of tank forces the old linear warfare is replaced by circular warfare.

B.H. Liddell Hart, *Thoughts on War*, 1944

Technology

The aim of military study should be to maintain a close watch upon the latest technical, scientific, and political developments, fortified by a sure grasp of the eternal principles upon which the great captains have based their contemporary methods, and inspired by a desire to be ahead of any rival army in securing options in the future.

B.H. Liddell Hart, *Thoughts on War*, 1944

Tempo

A commander must accustom his staff to a high tempo from the outset and continually keep them up to it. If he once allows himself to be satisfied with norms, or anything less than an all-out effort, he gives up the race from the starting-post and will sooner or later be taught a bitter lesson by his faster-moving enemy and be forced to jettison all his fixed ideas.

Field Marshal Erwin Rommel, 1891–1944, in B.H. Liddell Hart (ed.), *The Rommel Papers*.

Terrain

Those who do not know the conditions of mountains and forests, hazardous defiles, marshes and swamps, cannot conduct the march of any army. Those who do not use native guides are unable to obtain the advantages of the ground.

Sun Tzu, 400–320 BC, *The Art of War*, Chapter 11

In peace, soldiers must learn the nature of the land, how steep the mountains are, how the valleys debouch, where the plains lie, and understand the nature of rivers and swamps – then by means of the knowledge and experience gained in one locality, one can easily understand any other that it may be necessary to observe.

Niccolo Machiavelli, *The Art of War*, 1520

There is in every battlefield a decisive point the possession of which, more than any other, helps to secure victory by enabling its holder to make a proper application of the principles of war.

> Antoine Henri Jomini, *Summary of the Art of War*, 1838

Territorial Soldier

No militia will ever acquire the habits necessary to resist a regular force ... The firmness requisite for the real business of fighting is only to be attained by a constant course of discipline and service. I have never yet been witness to a single instance that can justify a different opinion, and it is most earnestly to be wished that the liberties of America may no longer be trusted in any material degree, to so precarious a dependence.

> George Washington, 1732–99

Terrorism

We have seen in the last few years the growth of a cult of political violence preached and practised not so much between states as within them. It is a sombre thought but it may be that in the 1970s civil war, rather than war between nations, will be the main danger that we face.

> Edward Heath, 1971, quoted in M. Dewar, *Weapons and Equipment of Counter Terrorism*, 1988.

Time

In military operations, time is everything.

> The Duke of Wellington, despatch, 30 June 1800

Time is everything; five minutes makes the difference between victory and defeat.

> Horatio Nelson, 1758–1805

Go, sir, gallop, and don't forget that the world was made in six days. You can ask me for anything you like, except time.

> Napoleon I, to a staff officer, 1803

Our cards were speed and time, not hitting power, and these gave us strategical rather than tactical strength. Range is more to strategy than force.

> T.E. Lawrence, 'Guerrilla Warfare', *Encylopaedia Britannica*, 1929

Tradition

Every trifle, every tag or ribbon that tradition may have associated with the former glories of a regiment should be retained, so long as its retention does not interfere with efficiency.

> Colonel Clifford Walton, *History of the British Standing Army, 1660–1700*, 1894

The value of tradition to the social body is immense. The veneration for practices, or for authority, consecrated by long acceptance, has a reserve of strength which cannot be obtained by any novel device.

> Alfred Thayer Mahan, 1840–1914, *The Military Rule of Obedience*

It takes the Navy three years to build a ship. It would take three hundred to rebuild a tradition.

> Sir Andrew Browne Cunningham, to his staff, disapproving the recommendations that the Royal Navy save its ships by retiring from Crete and abandoning the soldiers ashore, May 1941

The spirit of discipline, as distinct from its outward and visible guises, is the result of association with martial traditions and their living embodiment.

> B.H. Liddell Hart, *Thoughts on War*, 1944

Don't talk to me about naval tradition. It's nothing but rum, sodomy and the lash.

> Sir Winston Churchill, 1874–1965, quoted in 'Former Naval Person' (date and author unknown)

Fortune is rightly malignant to those who break with the traditions and customs of the past.

> Sir Winston Churchill, note to the Foreign Secretary, 23 April 1945

This modern tendency to scorn and ignore tradition and to sacrifice it to administrative convenience is one that wise men will resist in all branches of life, but more especially in our military life.

> Sir Archibald Wavell, address to the officers of the Canadian Black Watch, Montreal, 1949

However praiseworthy it may be to uphold tradition in the field of soldierly ethics, it is to be resisted in the field of military command.

> Field Marshal Erwin Rommel, *The Rommel Papers*, 1953

Trafalgar (21 October 1805)

The signal has been made that the Enemy's combined fleet are coming out of port ... May God Almighty give us success over these fellows and enable us to get a Peace.

> Horatio Nelson, last letter to Lady Hamilton, unfinished, 19 October 1805 delivered after Nelson's death

Trafalgar was not only the greatest naval victory, it was the greatest and most momentous victory won either by land or by sea during the whole of the revolutionary War. No victory, and no series of victories, of Napoleon produced the same effect upon Europe ... Nelson's last triumph left England in such a position that no means remained to injure her.

> Charles Alan Fyffe, *History of Modern Europe*, 1883

Training

What is necessary to be performed in the heat of action should be practised in the leisure of peace.
> Vegetius, *The Military Institutions of the Romans*, AD 378

The main end and design of all the care and pains that are bestowed in keeping up good order and discipline is to fit and prepare an army to engage in a proper manner.
> Niccolo Machiavelli, *The Art of War*, 1520

The troops should be exercised frequently, cavalry as well as infantry, and the general should often be present to praise some, to criticize others, and to see with his own eyes that the orders ... are observed exactly.
> Frederick the Great, *Instructions for His Generals*, 1747

A good general, a well organized system, good instructions, and severe discipline, aided by effective establishments, will always make good troops, independently of the cause for which they fight. At the same time, a love of country, spirit of enthusiasm, a sense of national honour, and fanaticism will operate upon young soldiers with advantage.
> Napoleon I, *Maxims of War*, 1831

In no other profession are the penalties for employing untrained personnel so appalling or so irrevocable as in the military.
> General Douglas MacArthur, Annual Report as Chief of Staff US Army, 1933

If the exercise is subsequently discussed in the officers' mess, it is probably worth while; if there is argument over it in the sergeants' mess, it is a good exercise; while if it should be mentioned in the corporals' room, it is an undoubted success.
> Sir Archibald Wavell, in *Journal of the Royal United Services Institution*, May 1933

Train in difficult, trackless, wooded terrain. War makes extremely heavy demands on the soldier's strength and nerves. For this reason make heavy demands on your men in peacetime.

Field Marshal Erwin Rommel, *Infantry Attacks*, 1937

The safety and honour of Britain depend not on her wealth and administration, but on the character of her people. This in turn depends on the institutions which form character. In war, it depends, in particular, on the military institutions which create the martial habits of discipline, courage, loyalty, pride and endurance.

Sir Arthur Bryant, 'The Fate of the Regiment', in the *Sunday Times*, 4 April 1948

The best form of 'welfare' for the troops is first-class training.

Field Marshal Erwin Rommel, *The Rommel Papers*, 1953

Trenches

When I hear talk of lines, I always think I am hearing talk of the walls of China. The good ones are those that nature has made, and the good entrenchments are good dispositions and brave soldiers.

Maurice de Saxe, *Reveries*, 1732

To bury an army in entrenchment, where it may be outflanked and surrounded or forced in front even if secure from a flank attack, is manifest folly.

Antoine Henri Jomini, *Summary of the Art of War*, 1838

I knew a simple soldier boy
Who grinned at life in empty joy,
Slept soundly through the lonesome dark,
And whistled early with the lark.

In winter trenches, cowed and glum,
With cramps and lice and lack of rum.
He put a bullet through his brain.
No one spoke of him again.

Siegfried Sassoon, *Pray God*, 1918

Trumpet

And it shall come to pass, that when they make a long blast with the ram's horn, and when ye hear the sound of the trumpet, all the people shall shout with a great shout and the wall of the city shall fall down flat.

Joshua 6

With harsh-resounding trumpets' dreadful bray ...

William Shakespeare, *King Richard II*, 1595

Make all our trumpets speak; give them all breath,
Those clamorous harbingers of blood and death.

William Shakespeare, *Macbeth*, 1605

So he passed over, and all the trumpets sounded for him on the other side.

John Bunyan, *The Pilgrim's Progress*, 1678

Turnout

Smartness is the cement, but not the bricks.

B.H. Liddell Hart, *Thoughts on War*, 1944

We found it a great mistake to belittle the importance of smartness in turn-out, alertness of carriage, cleanliness of person, saluting, or precision of movement, and to dismiss them as naive, unintelligent, parade-ground stuff. I do not believe that troops can have unshakeable battle discipline without showing those outward and formal signs which mark the pride men take in themselves and their units and the mutual confidence and respect that exists between them and their officers.

Sir William Slim, *Defeat into Victory*, 1956

Uniform

I think it indifferent how a soldier is clothed, provided it is in a uniform manner; and that he is forced to keep himself clean and smart, as a soldier ought to be.

> The Duke of Wellington, letter to the War Office from Portugal, 1811

A soldier must learn to love his profession, must look to it to satisfy all his tastes and his sense of honour. That is why handsome uniforms are useful.

> Napoleon I, to General Gaspard Gourgaud, St Helena, 1815

The better you dress a soldier, the more highly will he be thought of by women.

> Field Marshal Lord Wolseley, *The Soldier's Pocketbook*, 1869

The secret of uniform was to make a crowd solid, dignified, impersonal; to give it the singleness and tautness of an upstanding man.

> T.E. Lawrence, *Revolt in the Desert*, 1927

This death's livery which walled its bearers from ordinary life, was sign that they had sold their wills and bodies to the State.

> T.E. Lawrence, 1888–1935

Valour

The better part of valour is discretion.
> William Shakespeare, *I King Henry IV*, 1597

For Valour.
> Inscription on the Victoria Cross, instituted 29
> January 1856

Victory

Know the enemy, know yourself; your victory will never be endangered. Know the ground, know the weather; your victory will then be total.
> Sun Tzu, 400–320 BC, *The Art of War*, Chapter 10.

A skilled commander seeks victory from the situation, and does not demand it from his subordinates.
> Sun Tzu, 400–320 BC, *The Art of War*

Victory in war does not depend entirely upon numbers or mere courage; only skill and discipline will insure it.
> Vegetius, *The Military Institutions of the Romans*, i, AD
> 378

Now are our brows bound with victorious wreaths,
Our bruised arms hung up for monuments.
> William Shakespeare, *King Richard III*, 1592

The enemy came. He was beaten. I am tired. Good night.
> Henri de la Tour d'Auvergne Turenne, after the battle
> of Tünen, 14 June 1658

A victory is very essential to England at the moment.
> Sir John Jervis (Lord St Vincent): before the Battle of
> Cape St Vincent, 14 February 1797

Madam, there is nothing so dreadful as a great victory –
excepting a great defeat.

> Attributed to the Duke of Wellington, 1769–1852

I do not deserve more than half the credit for the battles I have
won. Soldiers generally win battles; generals get credit for
them.

> Napoleon I, to General Gaspard Gourgaud,
> St Helena, 1818

Man does not enter battle to fight, but for victory. He does
everything he can to avoid the first and obtain the second.

> Ardant du Picq, 1821–70, *Battle Studies*

The will to conquer is victory's first condition, and therefore
every soldier's first duty.

> Ferdinand Foch, *Principles of War*, 1920

Victory at all costs, victory in spite of all terror, victory however
long and hard the road may be; for without victory there is no
survival.

> Winston Churchill, to the House of Commons, 13
> May 1940

Gaining military victory is not in itself equivalent to gaining the
object of war.

> B.H. Liddell Hart, *Thoughts on War*, 1944

Volunteer

Our voluntary service regulars are the last descendants of
those rulers of the ancient world, the Roman legionaries.

> Sir Ian Hamilton, *Gallipoli Diary*, 1920

War

War is a matter of vital importance to the state; the province of life or death; the road to survival or ruin.
> Sun Tzu, 400–320 BC, *The Art of War*, chapter 1

There has never been a protracted war from which a country has benefited.
> Sun Tzu, 400–320 BC, *The Art of War*, chapter 2

In war trivial causes produce momentous events.
> Julius Caesar, *The Gallic War*, i, 51 BC

Wars are the dread of mothers. (Bella detestata matribus.)
> Horace, *Odes*, i, *c.* 20 BC

The fear of war is worse than war itself.
> Seneca, *Hercules Furens*, *c.* AD 50

A necessary war is a just war.
> Niccolo Machiavelli, *The Prince*, 1513

War is the greatest plague that can afflict humanity; it destroys religion, it destroys states, it destroys families. Any scourge is preferable to it.
> Martin Luther: *Table-Talk*, 1569

Grim-visaged War hath smoothed his wrinkled front.
> William Shakespeare, *King Richard III*, 1592

... To reap the harvest of perpetual peace
By this one bloody trial of sharp war.
> William Shakespeare, *King Richard III*, 1592

Farewell the plumed troop, and the big wars,
That make ambition virtue! O, farewell!
Farewell the neighing steed, and the shrill trump,
The spirit-stirring drum, the ear-piercing fife,
The royal banner and all quality,
Pride, pomp and circumstance of glorious war!
> William Shakespeare, *Othello*, 1604

A civil war is like the heat of fever; but a foreign war is like the heat of exercise, and serveth to keep the body in health.
> Francis Bacon, *Essays*, 1625

Every man is bound by nature, as much as in him lieth, to protect in war the authority by which he is himself protected in time of peace.
> Thomas Hobbes, *Leviathan* (conclusion), 1651

War is the trade of Kings.
> John Dryden, *King Arthur*, 1691

War is the best academy in the world, where men study by necessity and practice by force, and both to some purpose, with duty in the action, and a reward in the end.
> Daniel Defoe, *Essay upon Projects*, 1692

War! that mad game the world so loves to play.
> Jonathan Swift, 1667–1745

I have loved war too well.
> Louis XIV of France, on his deathbed, 1715

War is a science replete with shadows in whose obscurity one cannot move with assured step. Routine and prejudice, the natural result of ignorance, are its foundation and support. All sciences have principles and rules. War has none. The great captains who have written of it give us none. Extreme cleverness is required merely to understand them.
> Maurice de Saxe, *Reveries*, 1732

The circumstances of war are sensed rather than explained.
> Maurice de Saxe, 1696–1750, *Letters and Memoirs*.

Its the fashion now to make war and presumably it will last a good long while.

> Frederick the Great, letter to Voltaire, 1742

Even war is pusillanimously carried on in this degenerate age; quarter is given; towns are taken, and the people spared; even in a storm, a woman can hardly hope for the benefit of a rape.

> Lord Chesterfield, letter to his son, 12 January 1757

War is in its nature hazardous and an option of difficulties.

> General Wolfe, letter to Captain William Rickson,
> 5 November 1757

The only way to save out empires from the encroachment of the people is to engage in war, and thus substitute national passions for social aspirations.

> Attributed to Catherine of Russia, 1729–96

The art of war is the most difficult of all arts; therefore military glory is universally considered the highest, and the services of warriors are rewarded by a sensible government in a splendid manner and above all other services.

> Napoleon I, *Maxims of War*, 1831

There is more of misery inflicted upon mankind by one year of war than by all the civil peculations and oppressions in a century. Yet it is a state into which the mass of mankind rush with a greatest avidity, hailing official murderers, in scarlet and gold, and cock's feathers, as the greatest and most glorious of human creatures.

> Sydney Smith, in the *Edinburgh Review*, 1813

War is the foundation of all the arts, because it is the foundation of all the high virtues and faculties of men.

> John Ruskin, 1819–1900

War ought never to be undertaken but under circumstances which render all intercourse of courtesy between the combatants impossible. It is a bad thing that men should hate each other; but it is far worse that they should contract the habit of cutting one another's throats without hatred. War is never lenient but where it is wanton; when men are compelled to fight in self-defence, they must hate and avenge; this may be bad; but it is human nature.

> T.B. Macaulay, *On Milford's History of Greece*, 1824

To carry the spirit of peace into war is a weak and cruel policy. When an extreme case calls for that remedy which is in its own nature most violent, and which in such cases, is a remedy only because it is violent, it is idle to think of mitigating and diluting. Languid war can do nothing which negotiation or submission will not do better: and to act on any other principle is not to save blood and money, but to squander them.

T.B. Macaulay, *Hallam*, 1828

The conduct of war resembles the workings of an intricate machine with tremendous friction, so that combinations which are easily planned on paper can be executed only with effort.

Karl von Clausewitz, *Principles of War*, 1832

War is an act of violence pushed to its utmost limits.

Karl von Clausewitz, *On War*, 1832

War is an act of violence whose object is to constrain the enemy to accomplish our will.

Karl von Clausewitz, *On War*, 1832

Der Krieg ist nichts als eine Fortsetzung der politischen verkehrs mit Einmischung anderer Mittel. (Commonly translated as 'war is the continuation of policy by other means.')

Karl von Clausewitz, *On War*, 1832

War admittedly has its own grammar, but not is own logic.

Karl von Clausewitz, *On War*, 1832

War is not only chameleon-like in character, because it changes its colours in some degree in each particular case, but it is also, as a whole, in relation to the predominant tendencies that are in it, a wonderful trinity, composed of the original violence of its elements, hatred and animosity, which may be looked upon as blind instinct; of the play of probabilities and chance, which make it a free activity of the soul; and of the subordinate nature of a political instrument by which it belongs purely to reason.

Karl von Clausewitz, *On War*, 1832

The most just war is one which is founded upon undoubted rights and which, in addition, promises to the state advantages commensurate with the sacrifices required and the hazards incurred.

Antoine Henri Jomini, *Summary of the Art of War*, 1838

The difference of race is one of the reasons why I fear war may always exist; because race implies difference, difference implies superiority, and superiority leads to predominance.

> Benjamin Disraeli, to the House of Commons,
> 1 February 1849

Neither war nor anything else can change in its essentials. If it appears to do so it is because we are still mistaking accidents for essentials.

> Sir Julian Corbett, *The Successors of Drake*, 1900

It is magnificent, but it is not war.

> Pierre Bosquet, on observing the charge of the Light
> Brigade, Balaklava, 25 October 1854

I believe that war is at present productive of good more than of evil.

> John Ruskin, *Modern Painters*, 1856

If there be greater calamity to human nature than famine, it is that of an exterminating war.

> Benjamin Disraeli, speech, Mansion House, 9
> November 1877

War makes the victor stupid and the vanquished vengeful.

> F.W. Nietzsche, *Human All-Too-Human*, 1878

The man who has renounced war has renounced a grand life.

> F.W. Nietzsche, *The Twilight of the Idols*, 1889

The success of a war is gauged by the amount of damage it does.

> Victor Hugo, *Ninety-Three*, 1879

Der Krieg ein Glied in Gottes Weltordnung ... Ohne den Krieg würde die Welt in Materialismus versumpfen. (War is a necessary part of God's arrangement of the world ... Without war, the world would slide dissolutely into materialism.)

> Helmuth von Moltke ('The Elder'), letter to J.K.
> Bluntschili, 11 December 1880

Eternal peace is a dream, and not even a beautiful one ... In [war] are developed the noblest virtues of man: courage and abnegation, dutifulness and self-sacrifice.

> Helmuth von Moltke ('The Elder'), letter to J.K. Bluntschili, 11 December 1880

To abolish war we must remove its cause, which lies in the imperfection of human nature.

> Colmar von der Goltz, *The Nation in Arms*, 1883

War puts nations to the Test. Just as mummies fall to pieces the moment they are exposed to air, so war pronounces its sentence of death on those social institutions which have become ossified.

> Karl Marx, 1818–83, in *Sochineniya*

All the business of war, and indeed all the business of life, is to endeavour to find out what you don't know by what you do; that's what I called 'guessing what was at the other side of the hill'.

> The Duke of Wellington, quoted in *Croker Papers*, vol. iii, 1885

War, with its many acknowledged sufferings, is above all harmful when it cuts a nation off from others and throws it back upon itself.

> Alfred Thayer Mahan, *The Influence of Sea Power upon History*, 1890

As long as war is regarded as wicked it will always have its fascinations. When it is looked upon as vulgar, it will cease to be popular.

> Oscar Wilde, *Intentions*, 1891

... the coward's art of attacking mercilessly when you are strong, and keeping out of harm's way when you are weak. That is the whole secret of successful fighting. Get your enemy at a disadvantage; and never, on any account, fight him on equal terms.

> George Bernard Shaw, *Arms and the Man*, 1894

Oh, war! war! the dream of patriots and heroes! A fraud. A hollow sham, like love.

> George Bernard Shaw, *Arms and the Man*, 1894

God will see to it that war shall always recur, as a drastic medicine for ailing humanity.

> Heinrich von Treitschke, *Politics*, 1897

War makes rattling good history; but peace is poor reading.

> Thomas Hardy, *The Dynasts*, 1906

War is a dreadful thing, and unjust war is a crime against humanity. But it is such a crime because it is unjust, not because it is war.

> Theodore Roosevelt, speech at the Sorbonne, 23 April 1910

War is a biological necessity of the first importance.

> Friedrich von Bernhardi, *Germany and the Next War*, 1911

The essence of war is violence; moderation in war is imbecility.

> Sir John Fisher, letter to Lord Esher, 25 April 1912

Until the world comes to an end the ultimate decision will rest with the sword.

> Wilhelm II of Germany, speech in Berlin, 1913

The lamps are going out all over Europe; we shall not see them lit again in our lifetime.

> Sir Edward Grey, on the evening of 4 August 1914, when war with Germany was inevitable

... the struggle that will decide the course of history for the next hundred years.

> Helmuth von Moltke ('The Younger'), letter to Field Marshal Conrad von Hötzendorff, 5 August 1914

For all we have and are,
For all our children's fate,
Stand up and take the war,
The Hun is at the gate!

> Rudyard Kipling, *For All We Have and Are*, 1914

I have seen war, and faced modern artillery, and I know what an outrage it is against simple men.

> T.M. Kette, *The Ways of War*, 1915

There are poets and writers who see naught in war but carrion, filth, savagery, and horror ... They refuse war the credit of being the only exercise in devotion on the large scale existing in this world. The superb moral victory over death leaves them cold. Each one to his taste. To me this is no valley of death – it is a valley brim full of life at its highest power.

> Sir Ian Hamilton, 30 May 1915, *Gallipoli Diary*, 1920

As I reflected upon the intensive application of man to war in cold, rain and mud; in rivers, canals, and lakes; underground and in the air, and under the sea; infected with vermin, covered with scabs, adding the stench to his own filthy body to that of his decomposing comrades; hairy, begrimed, bedraggled, yet with unflagging zeal striving eagerly to kill his fellows; and as I felt within myself the mystical urge of the sound of great cannon, I realized that war is a normal style of man.

> G.W. Crile, *A Mechanistic View of War and Peace*, 1915

Generals cannot be entrusted with anything – not even with war.

> Attributed to Georges Clemençeau, 1841–1929 (often quoted as 'War is too important to be left to the generals'). Also attributed to Talleyrand and Briand.

Ah, God, sweet is war, with its songs, with its prolonged leisures.

> Guillaume Appolinaire, 1880–1918

Above all, this book is not concerned with Poetry,
The subject of it is War, and the Pity of War,
The Poetry is in the pity.

> Wilfred Owen, *Poems*, preface, 1920

The war to end wars has resulted in a peace to end peace.

> Attributed to Kaiser Wilhelm II, on being appraised of the terms of the Treaty of Versailles, June 1919

War is the highest expression of the racial will to life.

> Erich Ludendorff: *My War Memoirs*, 1919

In war, only what is simple can succeed.

> Paul von Hindenburg, *Out of My Life*, 1920

... a Rabelaisian game of chess where the board has a million squares and the pieces consist of a dozen Kings and Queens, a thousand Knights, and so many pawns that no one can exactly count them.

Sir Ian Hamilton, *The Soul and Body of an Army*, 1921

Once in a generation, a mysterious wish for war passes through a people. Their instinct tells them there is no other way of progress and of escape from habits that no longer fit them. Whole generations of statesmen will fumble over reforms for a lifetime which are put into full-blooded execution within a week of a declaration of war. There is no other way. Only by intense sufferings can the nations grow, just as a snake once a year must with anguish slough off the once beautiful coat which has now become a strait jacket.

Sir Ian Hamilton, *Gallipoli Diary*, 1920

There is nothing certain about war except that one side won't win.

Sir Ian Hamilton, *Gallipoli Diary*, 1920

War is the continuation of the policy of peace; peace is the continuation of the policies of war.

V.I. Lenin, 1870–1924, in *Polnoe Sobranie Sochineniya*

Any distinction between belligerents and non-belligerents is no longer admissible today either in fact or theory ... When nations are at war, everyone takes part in it; the soldier carrying his gun, the woman loading shells at a factory, the farmer growing wheat, the scientist experimenting in his laboratory ... It begins to look now as if the safest place may be the trenches.

Giulio Douhet, *The Command of the Air*, 1921

War is a simple art: its essence lies in its accomplishment.

Ferdinand Foch, *Precepts* 1919

The military mind always imagines that the next war will be on the same lines as the last. That has never been the case and never will be.

Ferdinand Foch, *Precepts* 1919

There is no 'science' of war, and there never will be any. There are many sciences war is concerned with. But war itself is not a science; war is practical art and skill.

> Leon Trotsky, *How the Revolution Developed its Military Power*, 1924

War alone brings up to its highest tension all human energy and puts the stamp of nobility upon the people who have the courage to face it.

> Benito Mussolini, 1883–1945, article in *The Italian Encyclopaedia*

There is such a horror of war in the great nations who passed through Armageddon that any declaration or public speech against armaments, although it consisted only of platitudes and unrealities, has always been applauded; and any speech or assertion that set forth the blunt truth has been incontinently relegated to the category of 'war monger' … The cause of disarmament will not be obtained by mush, slush and gush. It will be advanced steadily by the harassing expense of fleets and armies and by the growth in confidence in a long peace.

> Sir Winston Churchill, 1932

There is no better teacher of war than war.

> Mao Tse-tung, *On the Study of War*, 1936

We are for the abolition of war, we do not want war; but war can only be abolished through war, and to get rid of the gun, we must first grasp it in our own hands.

> Mao Tse-tung, *Problems of War and Strategy*, 1938

I have seen war. I have seen war on land and sea. I have seen blood running from the wounded. I have seen men coughing out their gassed lungs. I have seen the dead in the mud. I have seen cities destroyed. I have seen two hundred limping, exhausted men come out of the line – the survivors of a regiment of a thousand that went forward 48 hours before. I have seen children starving. I have seen the agony of mothers and wives. I hate war.

> Franklin D. Roosevelt, speech at Chatauqua, New York, 14 August 1936

In war, whichever side may call itself the victor, there are no winners, only losers.

> Neville Chamberlain, speech at Kettering, England, 2 July 1938

There has never been a war yet which, if the facts had been put calmly before the ordinary folk, could not have been prevented. The common man is the greatest protection against war.

> Ernest Bevin, to the House of Commons, November 1945

The main ethical objection to war for intelligent people is that it is so deplorably dull and usually so inefficiently run ... Most people seeing the muddle of war forget the muddles of peace and the general efficiency of the human race in ordering its affairs.

> Sir Archibald Wavell, unpublished 'Recollections', 1947

War, which used to be cruel and magnificent, has now become cruel and squalid. It is all the fault of democracy and science. From the moment that either of these meddlers and muddlers was allowed to take part in actual fighting, the doom of War was sealed. Instead of a small number of well-trained professionals championing their country's cause with ancient weapons and a beautiful intricacy of archaic movement, we now have entire populations, including even women and children, pitted against each other in brutish mutual extermination, and only a set of blear-eyed clerks left to add up the butcher's bill. From the moment when Democracy was admitted to, or rather forced itself upon, the battlefield, War ceased to be a gentlemen's pursuit.

> Sir Winston Churchill, *My Early Life*, 1930

In war, resolution; in defeat, defiance; in victory, magnanimity; in peace, goodwill.

> Epigram after the Great War, 1914–18. Sir Winston Churchill, quoted by Sir Edward Marsh in *A Number of People*, 1939. Later used as the 'Moral of the Work' in each volume of Churchill's *The Second World War*

Death and sorrow will be the companions of our journey; hardship our garment; constancy and valour our only shield. We must be united, we must be undaunted, we must be inflexible.

> Sir Winston Churchill, to the House of Commons, 8 October 1940

I have used one principle in these operations:
  'Fill the unforgiving minute
  With sixty seconds worth of distance run'
That is the whole art of war, and when you get to be a general, remember it.

> George Patton, quoting Kipling, France, 28 August 1944

In mortal war, anger must be subordinated to defeating the main immediate enemy.

> Sir Winston Churchill, *The Gathering Storm*, 1948

There is no merit in putting off a war for a year if, when it comes, it is a far worse war or one much harder to win.

> Sir Winston Churchill, *The Gathering Storm*, 1948

To jaw-jaw is better than to war-war.

> Sir Winston Churchill, Washington, 26 June 1954

Whereas the other arts are, at their height, individual, the art of war is essentially orchestrated.

> B.H. Liddell Hart, *Thoughts on War*, 1944

War is always a matter of doing evil in the hope that good may come of it.

> B.H. Liddell Hart, *Defence of the West*, 1950

Unlike mathematics, war is an empirical matter. War is history; this means that its laws are deductions to be made only after the event.

> Jean Dutourd, *Taxis of the Marne*, 1957

Thus war, the horseman, turned back to his crimson courts and dragged brave gallants by their belts, girls by their braids, and hung small children from his saddle-horns in clusters. Behind him the blind followed, stumbling with long staffs, and some way back the cripples, the armless, the half-wits, and mothers in long rows who walked alive toward Hades.

> Nikos Kazantzakis, *The Modern Odyssey, a Sequel*, 1958

Future generations may discuss the Second World War as 'just another war'. Those who experienced it know that it was a war justified in its aims and successful in accomplishing them. Despite all the killing and destruction that accompanied it, the Second World War was a good war.

> A.J.P. Taylor, *The Second World War*, 1975

Mankind must put an end to war – or war will put an end to mankind.

> President John F. Kennedy, to the General Assembly of the United Nations, 25 September 1961

War, Principles of

War remains an art and, like all arts whatever its variation, will have its ending principles. Many men, skilled either with sword or pen and sometimes with both, have tried to expound those principles. I heard them once from a soldier of experience for whom I had a deep and well-founded respect. Many years ago, as a cadet hoping some day to be an officer, I was poring over 'The Principles of War', listed in the old Field Services Regulations, when the Sergeant-Major came upon me. He surveyed me with kindly amusement. 'Don't bother your head about all them things, me lad,' he said. 'There's only one principle of war and that's this. Hit the other fellow as quick as you can and as hard as you can, where it hurts him most, when he ain't looking.'

> Sir William Slim, *Defeat into Victory*

The US has broken the second rule of war. That is, don't go fighting with your land army on the mainland of Asia. Rule

One is don't march on Moscow. I developed these two rules myself.

Montgomery of Alamein, quoted in Chalfont's
Montgomery of Alamein, 1976

Warship

A man-of-war is the best ambassador.

Oliver Cromwell, 1599–1658

Taken all in all, a ship of the line is the most honourable thing that man, as a gregarious animal, has ever produced.

John Ruskin, *The Harbours of England*, 1856

Waterloo

Would God that night or Blücher would come.

Attributed to Wellington, on the afternoon of
Waterloo, 18 June 1815

It has been a damned serious business – Blücher and I have lost 30,000 men. It has been a damned nice thing – the nearest run thing you ever saw in your life ... By God! I don't think it would have done if I had not been there.

The Duke of Wellington to Thomas Creevey at
Brussels, the day after Waterloo, 19 June 1815

You will have heard of our battle of the 18th. Never did I see such a pounding match ... Napoleon did not manoeuvre at all. He just moved forward in the old style, and was driven off in the old style.

The Duke of Wellington, letter to Sir William
Beresford, 2 July 1815

Meeting an acquaintance of another regiment, a very little fellow, I asked him what had happened to them yesterday. 'I'll be hanged,' says he, 'if I know anything at all about the matter, for I was all day trodden in the mud and galloped over by every scoundrel who had a horse; and, in short, I only owe my existence to my insignificance!'

Captain John Kincaid, reminiscence of Waterloo in
Adventures with the Rifle Brigade, 1830

The battle of Waterloo was won on the playing fields of Eton.

Attributed to The Duke of Wellington. See Montalem-
bert, De l'avenir politique de l'Angleterre, 1856

Waterloo is a battle of the first rank won by a captain of the second.

Victor Hugo, Les Misérables, 1862

No incident is more familiar in our military history than the stubborn resistance of the British line at Waterloo. Through the long hours of the midsummer day, silent and immovable the squares and squadrons stood in the trampled corn, harassed by an almost incessant fire of cannon and of musketry, to which they were forbidden to make reply. Not a moment but heard some cry of agony; not a moment but some comrade fell headlong in the furrows. Yet as the bullets of the skirmishers hailed around them, and the great round shot tore through the tight-packed ranks, the word was passed quietly. 'Close in on the centre, men'; and as the sun neared his setting, the regiments, still shoulder to shoulder, stood fast upon the ground they had held at noon. The spectacle is characteristic. In good fortune and in ill it is rare indeed that a British regiment does not hold together; and this indestructible cohesion, best of all the qualities that an armed body can possess, is based not merely on hereditary resolution, but on mutual confidence and mutual respect. The man in the ranks has implicit faith in his officer, the officer an almost unbounded belief in the valour and discipline of his men; ...

Colonel Henderson, The Science of War, 1906

Wavell, Archibald Percival (1883–1950)

The only one who showed a touch of genius was Wavell.

Erwin Rommel, The Rommel Papers, 1953

Weapons

The means of destruction are approaching perfection with frightful rapidity.

> Antoine Henri Jomini, *Summary of the Art of War*, 1838

The instruments of battle are valuable only if one knows how to use them.

> Ardant du Picq, d. 1870, *Battle Studies*

The unresting progress of mankind causes continual change in the weapons; and with that must come a continual change in the manner of fighting.

> Alfred Thayer Mahan, *The Influence of Sea Power upon History*, 1890

Every development or improvement in firearms favours the defensive.

> Giulio Douhet, *The Command of the Air*, 1921

Every improvement in weapon power has aimed at lessening the danger on one side by increasing it on the other. Therefore, every improvement in weapons has eventually been met by a counter-improvement which has rendered the improvement obsolete, the evolutionary pendulum of weapon power, slowly or rapidly, swinging from the offensive to the protective and back again in harmony with the pace of civil progress, with each swing in a measurable degree eliminating danger.

> J.F.C. Fuller, *Armaments and History*, 1945

There are two universal and important weapons of the soldier which are often overlooked – the boot and the spade. Speed and length of marching has won many victories; the spade has saved many defeats and gained time for victory.

> Sir Archibald Wavell, *The Good Soldier*, 1945

Today the expenditure of billions of dollars every year on weapons, acquired for the purpose of making sure we never need to use them, is essential to keeping the peace.

> President John F. Kennedy, speech at American University, Washington, June 1963

Weather

Every mile is two in winter
>George Herbert, *Jacula Prudentum*, 1651

Russia has two generals whom she can trust – General Janvier and General Février.
>Tsar Nicholas I of Russia, as reported in *Punch*, 10 March 1853

Wellington, Sir Arthur Wellesley, Duke of, (1769–1852)

... a fine fellow with the best nerves of anyone I ever met with.
>Major General Sir Lowry Cole, letter from Portugal, 1811 (Cole was one of Wellington's divisional commanders.)

I should pronounce him to be a man of little genius, without generosity, and without greatness of soul.
>Napoleon I, letter to Barry O'Meara, St Helena, 20 September 1817

The sight of his long nose among us on a battle morning was worth ten thousand men, any day of the week.
>Captain John Kincaid, quoted in *Britain at Arms*, compiled by Thomas Gilby, 1963

Western Front

My God, did we really send men to fight in that?
>Lieutenant General Sir Launcelot Kiggell, on seeing the landscape after the battle of Passchendaele, 1917

The army report confined itself to a single sentence: All quiet on the Western Front.
>Erich Maria Remarque, *Nothing New in The West*, 1929

Soldiers of the Western Front, your hour has come. The fight which begins today will determine Germany's destiny for a thousand years.

> Adolf Hitler, Order of the Day, 10 May 1940

Wingate, Orde Charles (1903–44)

… a man of genius who might well have become also a man of destiny.

> Sir Winston Churchill, to the House of Commons, 2 August 1944

Withdrawal

The withdrawal should be thought of as an offensive instrument, and exercises be framed to teach how the enemy can be lured into a trap, closed by a counter-stroke or devastating circle of fire.

> B.H. Liddell Hart, *Thoughts on War*, 1944

Women

If upon service you have any ladies in your camp, be valiant in your conversation before them. There is nothing pleases the ladies more than to hear of storming breaches, attacking the covert-way sword in hand, and such like martial exploits.

> Francis Grose, *Advice to the Officers of the British Army*, 1782

Sir,
You, having thought fit to take to yourself a wife, are to look for no further attentions from your humble servant.

> Lord St Vincent, letter to Lieutenant Bayntum, 1795

… wants to go to Lisbon, and I have told him he may stay there 48 hours which is as long as any reasonable man can wish to stay in bed with the same woman.

> The Duke of Wellington, letter from Portugal, 1811

He who is full of courage and sang-froid before an enemy battery, amid the bullets, sometimes trembles and loses his head before a skirt or a peruke.

> Napoleon I, 1769–1821, conversation with Captain Poppleton, St Helena

In war, as in love, we must achieve contact ere we triumph.

> Napoleon I, *Political Aphorisms*, 1848

Marriage is good for nothing in the military profession.

> Napoleon I, *Political Aphorisms*, 1848

Wounds

A wound is nothing, be it ne'er so deep;
Blood is the god of war's rich livery.

> Christopher Marlowe, *Tamburlaine the Great*, 1587

A scar nobly got, or a noble scar, is a good livery of honour.

> William Shakespeare, *All's Well That Ends Well*, 1602

So well thy words become thee as thy wounds;
They smack of honour both.

> William Shakespeare, *Macbeth*, 1605

LORD UXBRIDGE. I've lost my leg, by God!
WELLINGTON. By God, sir, so you have!

> Conversation at Waterloo, 18 June 1815

'You're wounded!' 'Nay,' the soldier's pride
Touched to the quick, he said:
'I'm killed, Sire!' and his chief beside,
Smiling the boy fell dead.

> Robert Browning, *Incident of the French Camp*, 1846

Wise men took refuge in the virtues of cold water, and kept the surgeons at a safe distance.

> Sir John Fortescue, *History of the British Army*, 1899

Bibliography

Bond, Brian; Liddell Hart. *A Study of his Military Thought* (Cassell, 1977)

Churchill, Sir Winston, *The Second World War* (Cassell, 1948)
Vol.1: *The Gathering Storm*
Vol.2: *Their Finest Hour*
Vol.3: *The Grand Alliance*
Vol.4: *The Hinge of Fate*
Vol.5: *Closing the Ring*
Vol.6: *Triumph and Tragedy*

Clausewitz, Karl von, *On War: New and Revised Edition* (Keegan Paul Trench Taubner, 1908)

The Concise Oxford Dictionary of Quotations (Oxford University Press, 1981)

Dewar, Michael, *The Art of Deception in Wartime* (David and Charles, 1989)

Farrar-Hockley, A., *The Edge of the Sword* (Muller, 1954)

Fortescue, Sir John, *A Gallant Company* (Williams and Norgate, 1930s)

Fortescue, Sir John, *A History of the British Army* (Macmillan, 1920)

Fortescue, Sir John, *Following the Drum* (Blackwood, 1931)

Heinl, Robert, *Dictionary of Military and Naval Quotations* (United States Naval Institute, 1966)

Henderson, Colonel, *The Science of War* (Longman Green, 1906)

Kincaid, John, *Adventures with the Rifle Brigade* (T. and W. Boone, 1830)

Lawrence, T.E., *Seven Pillars of Wisdom* (Jonathan Cape, 1926)

Liddell Hart, B.H. (ed), *The Rommel Papers* (Collins, 1953)

Mahan, Alfred Thayer, *Influence of Seapower upon History* (Dover, 1988)

Majdalanay, Fred, *The Monastery* (Bodley Head, 1950)

Napoleon I, *Maxims* (5th edition, Paris, 1874)
Serve to Lead (HMSO, 1969)
Slim, Sir William, *Defeat into Victory* (Cassell, 1956)
Sun Tzu, *The Art of War: The Modern Chinese Interpretation* (David and Charles, 1987)
Turner, E.S., *Gallant Gentlemen* (Michael Joseph)

Numerous Leavenworth Papers, Combat Studies Institute, Fort Leavenworth, USA
Numerous Staff College Papers, Staff College, Camberley, UK
Private collections of quotations belonging to Ian Drury, Esq.
The author's private collection of quotations
Papers belonging to the author's grandfather, Vice-Admiral K.G.B. Dewar

Index

254

Shelley, Frances, Lady, 45
Sherman, W.T., 108, 189
Sidney, Philip, 58, 102
Skipton, Sergeant-Major-General Philip, 175
Slessor, John, 26, 27, 32, 100, 197
Slim, William, 23, 37, 39, 47, 74, 79, 94, 138, 139, 149, 151, 152, 153, 167, 180, 209, 227, 243
Smith, Brigadier General C.F., 46
Smith, Sydney, 233
Socrates, 114
Spee, Graf von, 99, 124
Spenser, Edmund, 50
Stalin, Joseph, 26, 34, 94
Stanhope, 154
Sun Tzu, 28, 70, 74, 77, 84, 90, 114, 116, 117, 135, 141, 186, 187, 202, 212, 213, 215, 218, 221, 229, 231
Sutcliffe, Matthew, 58
Swift, Jonathan, 32, 232

Tacitus, 50
Thackeray, W.M., 33, 52, 207
Talleyrand, Charles Maurice, 120
Tamerlane, 75
Tarle, E.V., 90
Tarleton, Banastre, 63
Taylor, A.J.P., 243
Taylor, Bayard, 52
Taylor, General Maxwell D., 104
Tennyson, Alfred Lord, 43, 45, 50, 53, 58, 82, 160, 195, 206
Terence, 51
Thoreau, Henry David, 45, 106
Thucydides, 60
Tiberius Sempronius Gracchus, 203
Tolstoy, Leo, 33
Torrington, Admiral Arthur, 110
Treitschke, Heinrich von, 237
Trotsky, Leon, 240
Tucker, Colonel J.G.D., 115
Turenne, Henri de la Tour d'Auvergne, 229
Turner, E.S., 189

Vegetius, 51, 91, 188, 202, 225, 229
Victoria, 86
Vincent, Lord St, 92, 105, 148, 160, 161, 163, 169, 199, 211, 248
Vinci, Leonardo da, 166
Virgil, 31
Voltaire, François-Marie Arouet, 22, 80, 105

Walpole, Horace, 126
Walton, Colonel Clifford, 223
Washington, George, 26, 54, 91, 169, 176, 190, 200, 222
Waterfield, Private Robert, 206
Wavell, Archibald, 60, 63, 74, 93, 118, 128, 129, 133, 143, 145, 150, 165, 172, 177, 187, 208, 224, 225, 241, 246
Webster, Daniel, 27
Wellington, Duke of, 20, 21, 32, 33, 35, 40, 45, 52, 59, 68, 92, 96, 97, 98, 113, 115, 117, 124, 126, 132, 133, 142, 144, 149, 154, 164, 176, 187, 190, 192, 194, 205, 206, 222, 228, 230, 236, 244, 245, 248
Wells, H.G., 150, 180
Wesley, John, 54
Whiting, William, 156
Wilde, Oscar, 236
Wilhelm I of Prussia, 93
Wilhelm II, Kaiser, 34, 217, 237, 238
Wilkinson, Spencer, 69
Willkie, Wendell, 55
Wilson, President Woodrow, 62
Windham, William, 29
Wingate, Orde, 215
Wintringham, Tom, 134
Wolfe, Charles, 55, 95, 120
Wolfe, James, 161, 184, 233
Wolseley, Field Marshal Lord, 95, 207, 228
Wu Ch' i, 73

Xenophon, 62, 71

Zuckmayer, Carl, 69

256